Salsebil Bel Hadj Ali

Mapping of hydro-morphological evolution and risk zones

Salsebil Bel Hadj Ali

Mapping of hydro-morphological evolution and risk zones

Contribution of remote sensing and hydraulic modeling

ScienciaScripts

Cover image: www.ingimage.com

This book is a translation from the original published under ISBN 978-613-8-40754-6.

Publisher:
Sciencia Scripts
is a trademark of
Dodo Books Indian Ocean Ltd. and OmniScriptum S.R.L publishing group

120 High Road, East Finchley, London, N2 9ED, United Kingdom
Str. Armeneasca 28/1, office 1, Chisinau MD-2012, Republic of Moldova, Europe
Printed at: see last page
ISBN: 978-620-6-00081-5

Contents

Acknowledgements

At the end of this work, I would like to express my sincere thanks to the people who helped me and who contributed to the elaboration of this report and to the success of these great academic years.

I would like to sincerely thank Dr. Fatma Trabelsi Bouchendira, who, as a supervisor, has always been attentive and very available throughout the realization of this work, as well as for the inspiration, the help and the time that she was willing to devote to me and without whom this thesis would never have seen the light.

My thanks also go to Professor Mohamed Raouf Mahjoub, Director General of the Higher School of Rural Equipment Engineers of Medjez Elbab (ESIER) for his generosity and great patience in spite of his academic and professional responsibilities. I would like to express my gratitude to all the consultants and internet users I met during my research who agreed to answer my questions with kindness, and in particular to Professor Christophe Zander, Professor of Social Sciences at the Institute of Social Work of Burgundy (IRTESS).

My sincere thanks also go to Colonel Lamine Aouni of the National Center for Remote Sensing and Mapping

I would like to express my gratitude to my aunt Moni who was there to support me throughout my academic career. I do not forget Yasser and Lina for their support and patience.

Finally, I address my most sincere thanks to all my relatives and friends.

Summary

The issue of natural hazards in general and Floods in particular is a topic of current interest that marks a memorable action in the world and specifically in Tunisia, especially in view of the last major floods of the Medjerda. Indeed, the management of this risk becomes more and more a necessity that must include all actors and all possible means available.

In this work, we have exposed the mapping of flood areas of the middle valley of the Medjerda by combining different complementary approaches of remote sensing coupled with GIS and hydraulic modeling (HEC RAS) and (HEC Geo RAS).

The processing and interpretation of Landsat 8 TM satellite images at different acquisition dates (2003, 2009 and 2011), allowed us to have a high resolution digital terrain model, a land use map at different dates and a morphological hydro map. Thus, simulations of three flood scenarios with three flows were made for different return periods of 5, 50 and 100 years.

The combination of all these results allowed us to elaborate the flood hazard map. This map seems to be one of the most effective means for an efficient management; it can be used as a basic document by the public authorities to define the general rules contributing to a better management of the urban space while constituting a means of information of the population on the risks of floods and a tool of organization to the decision-makers, and to them belongs the final choice of the strategy of fight against the flood risk.

Finally, an important descriptive and spatial database was created which allowed us to build a GIS - flooding of the middle valley of the Medjerda which can be used as a decision support tool for the authorities.

General introduction

Floods are the world's number one natural hazard. In recent decades, the damage caused by floods has been particularly destructive. The importance of this damage is mainly attributable to past urbanization and industrialization in flood plains (Torterotot, 1993), which have resulted in an increase in the vulnerability of property and people. The high stakes associated with flooding explain the efforts that are being made today to analyze and understand this phenomenon in order to reduce the risk.

Remote sensing, thanks to its synoptic vision, allows the study of vast geographical fields and constitutes a powerful tool for the study of the evolution of soil conditions. The processing of satellite data has become essential for the evaluation of natural resources and the mapping of surface conditions, hence its use in this study.

The study of changing soil conditions and land use is of interest to focus on environmental problems in general. It is necessary to determine the nature and mode of intervention of human communities that modify the overall land use patterns according to changing needs. Research and analysis of land use and occupancy provide a necessary information base for the planner, developer,...

It is within this framework is integrated the present study on the mapping of areas at risk of flooding of the middle valley of the Medjerda (Sidi Salem_Lâaroussia) through the interpretation of satellite images and hydraulic modeling. The objectives of this work are:

- The study of the morphological evolution of oued Medjerda between Sidi Salem and Lâaroussia

by the elaboration of a hydromorphological map.

- The development of the land use map
- The development of the flood hazard map by means of modeling

Hydraulics that will lead to the determination of the extent of floods and their impacts

- The elaboration of the flood risk map that will allow the evaluation of the

risks to property and people in this study area.

To meet these objectives, this brief is structured in three parts:

Part I. Flooding, remote sensing, and the study area.

Chapitre I. Flooding phenomenon.

Part I. Flooding, remote sensing, and study area.

Chapter I: Flooding Phenomenon

Introduction

In Tunisia, the flood phenomenon is old. Throughout history, there are dozens of times when regions have been affected. The best described and best known events, for the most part still in people's memory, are those recorded since the beginning of the last century and especially after the 1950s. The floods of 1969 (the whole country, including the center and north), 1973 (middle and lower Medjerda), 1982 (Sfax), 1990 (Sidi Bouzid region), 1995 (Tataouine), 2003 (Greater Tunis), 2007 (Sabbalet Ben Ammar), 2009 (Redayef) ... are all episodes that will mark for a long time the hydrological records of the country.

Are rainfall extremes becoming more and more recurrent, which would explain the major damage caused by floods over the last five or six decades? This question is difficult to answer, especially since, in most cases, we do not have sufficiently long chronicles to detect possible breaks in stationarity in the rainfall series. In view of the scientific uncertainty that hangs over the question of climate change in general, and especially its impact on rainfall trends, one element nevertheless appears certain: the hydrological changes inherent in excessive urbanization and the various, sometimes imprudent, development actions are constantly increasing the vulnerability of our cities and our spaces to the risk of flooding.

I. Definition of the flooding phenomenon

According to the definition given in the "Flood" information file (MEDD-PRIM): *"Flooding is a submersion, rapid or slow, of an area usually out of water".*

From the same file:

"In the broadest sense, floods include river overflows, rising water tables, runoff resulting from heavy rainstorms, flooding due to the failure of protective structures, estuarine floods resulting from the combination of high tides, low-pressure situations and river flooding.

Flooding can occur for a number of reasons, including:

(a) . **Overflow of a river**: The river leaves its bed and overflows on the major bed (slow flood in plain, torrential in mountain).

(b) . **Runoff** : Following intense rainfall, water runs off and rapidly concentrates in the watercourse, generating torrential floods. This phenomenon is related to the inability of the soil to infiltrate rainwater (rural runoff accompanied by mudflows, or urban runoff).

(c) . **Rising water tables** : The water rises, after one or more rainy years, through the water tables that outcrop producing a spontaneous flooding. This phenomenon particularly concerns low or poorly drained land.

(d) . **Marine flooding**: particular situation of coastal areas

Within the framework of this work, we limit ourselves to the case of floods which have for origin the overflow of a river. This case is the most common in Tunisia.

Indeed, the major element at the origin of a flood of plain is the fall of important precipitations on the catchment area.

1. The watershed and the genesis of flooding:

The watershed, or catchment area, is the essential element in any hydrological study associated with the topography of a given region. It is defined as a portion of the earth's surface within which the topographic slopes carry all runoff that occurs within it to a single outlet. The outlet of a watershed is the most downstream point in the river system through which all the water draining from the watershed passes. In a watershed, the topography, or set of slopes, defines the flow path and the organization of the drainage system, which depends on the water supply. To this end, the set of watercourses, permanent or temporary, which participate in the linear flow of the topographic surface represents the hydrological network which is one of the most important characteristics of the watershed and can take a multitude of forms. The diversity of the hydrographic network of a watershed is due to four main factors: the slope of the land, geology, climate and human presence (Xiaomin Che et *al.*, 2004)

Schematically, during heavy rainfall, part of the water infiltrates the soil, the rest runs off the slopes and is thus conveyed to the watercourses. When a very large quantity of water reaches the river, it overflows its "usual" bed (or minor bed) and thus gives rise to the phenomenon of flooding.

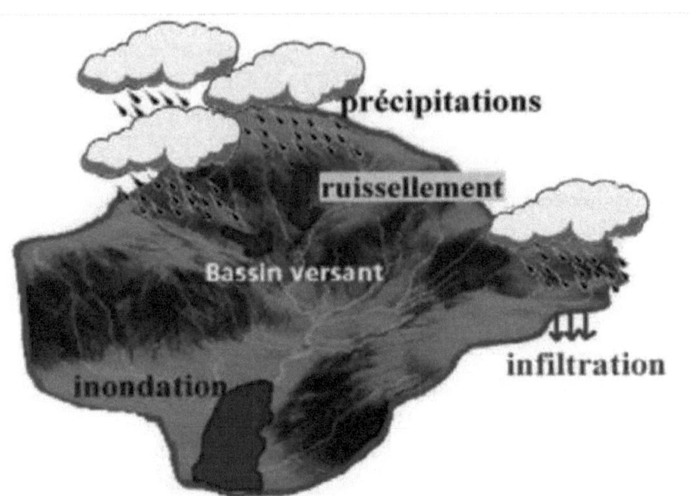

Figure 1. The genesis of the flooding phenomenon (river overflow). (Hostache, 2006)

2. The alluvial plain:

The term "alluvial plain" usually refers to the low-lying areas of the valley floor, made up of alluvial deposits made by the river during flooding. In terms of flow, the alluvial plain is often broken down into three zones: the minor bed, the middle bed and the major bed of the river. The minor bed corresponds to the flow area of the river except for overflow. The middle bed corresponds to the flow area for relatively low frequency floods. The major bed contains all the areas of the plain in which the river is likely to flow and overflow.

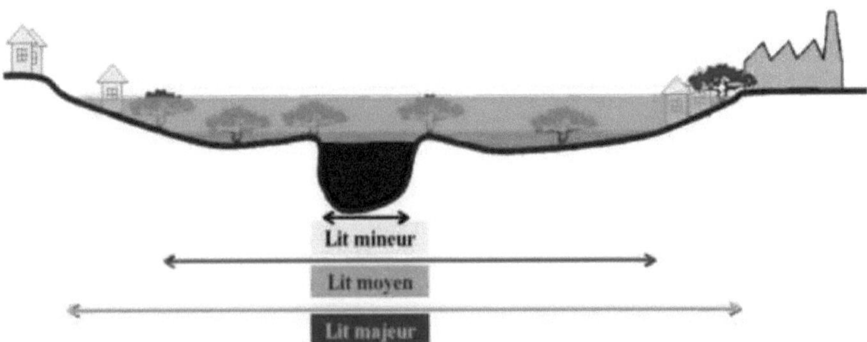

Figure 2. The alluvial plain. (Hostache, 2006)

II. Flood risk: hazards, issues and impacts :

The usual definition given for natural hazard is as follows:

(Risk) = (hazard) x (issue)

The risk is therefore the confrontation of a hazard (dangerous natural phenomenon) and a geographical area where there are human, economic or environmental issues.

1. Hazard

The hazard is a physical phenomenon, natural and uncontrollable, of given occurrence and intensity. It must be defined by an intensity (why and how?), a spatial occurrence (where?) and temporal (when?, duration?).

J The intensity reflects the importance of a phenomenon (Dauphiné, 2001). It can be measured (height of water for a flood, magnitude of an earthquake) or estimated (duration of submersion, speed of displacement).

J The probability of spatial occurrence is conditioned by predisposition or susceptibility factors (e.g. geological). The spatial extension of the hazard is more difficult to estimate (e.g. avalanche or ground movement).

J The temporal probability of occurrence depends on natural or anthropogenic triggers. It can be estimated qualitatively (negligible, weak, strong) or quantitatively (return period of 10 years, 30 years, 100 years).

The duration of the phenomenon must also be taken into account (duration considered for rainfall). It is often necessary to draw up a double entry table to characterize the hazard (intensity, duration).

2. Issues:

The notion of stake integrates people, goods and activities likely to be affected by the hazard. The consequences of the hazard on issues can be beneficial (groundwater recharge, supply of silt to agricultural land) or harmful (destruction of habitats, death of people, cutting of communication routes).

3. Flooding risks:

Risk represents the probability of material and economic damage, injury and/or death linked to the occurrence of a natural hazard (Ancey, 2005). It depends on a natural phenomenon obeying a law of probability, and on the stakes exposed in relation to the resources available to deal with it (Torterotot, 1993).

Risk characterization is a delicate issue. Its implementation on a catchment area consists in a hydrological modeling (Flow-Duration-Frequency), coupled with a hydraulic modeling and land use to obtain a cartographic representation of the risk.

Furthermore, mapping the extent of flooding is only a partial exploitation of satellite images. Indeed, as shown by some works (Horritt, 2000; Smith, 1997), satellite images of floods are rich in information, beyond the flood maps, which can be very beneficial to flood management and in particular to hydraulic modeling. For example, a fine three-dimensional characterization, coupled with hydraulic modeling, would allow a more complete exploitation of satellite images.

An increase in flood risk in an alluvial plain can be due to an increase in the stakes or an increase in the hazard. For example, urbanization is frequently responsible for an increase in risk for two reasons:

- The construction of houses in the alluvial plain increases the stakes.
- The construction of buildings, parking lots, roads, etc., makes part of the watershed more impermeable and leads to an increase in runoff, an increase in peak flow and a reduction in the time of concentration, resulting in an increase in the hazard and a reduction in the time available to deal with it.

1.1 **flooding impacts** :

a. Habitat Impacts:

Impacts on the habitat include everything related to the building and its contents as well as the people living in it. These damages are often divided into two categories:

- tangible damage: which can be evaluated in financial terms (impact on the load-bearing structure, on the finishing work, on the furniture, ...)
- intangible damages: which cannot be quantified in financial terms, such as impacts on people.

These two categories of damage are themselves divided into :

- direct damage: caused by the physical contact of water,
- consequential damages: due to temporary cessation of daily activities as a result of flooding.

b. Impacts on watershed morphology:

The main flood parameters on which the damage to the structure depends are:

- **the height of water**: the height of water exerts a pressure, static or dynamic, which can damage the surface layer of the soil.
- **the speed of the current**: to the hydrostatic pressure must be added a hydrodynamic pressure due to the speed of the current. High velocities can erode and dig into the soil.
- **the duration of immersion**: the longer the flooding lasts, the more the water rises by

capillary action causing the degradation of materials by swelling and hydrolysis.

- **floating bodies**: This type of flooding can be accompanied by mudflows. The debris and the speed at which it arrives depends on the particular conditions of the watercourse and the environment that can cause the narrowing of the wadi bed.

III. Flood modelling

A flood cannot be described from a single observation point or at a given time because it is a variable phenomenon in space and time. Classical hydrological observations are punctual and therefore localized.

Indeed, we will use in this study the one-dimensional hydraulic model HEC-RAS. It has a set of calculation possibilities allowing the elaboration of the water line. This model calculates the profiles of the water line from one cross-section to another while solving the following energy equation:

$$\frac{U^2}{2g}\alpha + z + h = \frac{U'^2}{2g}\alpha' + z' + h' + J$$

With : h and h' : draught at the level of sections 1 and 2.

z and z' : dimension of the bottom of the main channel.

U and U' : average velocity of the flow.

α and α': velocity weighting coefficients.

J : pressure drop between the two sections 1 and 2.

Indeed, the water line calculation procedure used by the HEC-RAS software is iterative, the main steps of which are :

4. Start from the water level at the upstream (or downstream) starting section.
5. Determine the total flow rate and the corresponding energy of the considered superposition.
6. Calculate the average pressure drop with the values from the previous step.
7. Solve the Bernouilli equation to determine the coastline of the water body.
8. Compare the calculated value and the value assumed in the first step and repeat these steps until the values agree to a user-defined tolerance (HEC-RAS user's Manual).

Conclusion

In the face of the scientific uncertainty that hangs over the issue of climate change in general, and especially its impact on rainfall trends, one element nevertheless appears certain: the hydrological changes inherent in the excessive urbanization and various development actions, sometimes imprudent, are constantly increasing the vulnerability of our cities and our spaces to the risk of flooding.

Chapter II: Theory of remote sensing

This chapter aims to describe and explain the general principles of passive optical and radar remote sensing techniques. It aims to allow an easier understanding of the following chapters in which satellite images will be used. In connection with the flooding problem, this chapter will be oriented towards the detection of water on satellite images.

Introduction

Remote sensing or remote sensing combines science, technology and art to acquire information about the earth's space without direct contact. The data are images acquired by a sensor on board a vehicle. The sensor records the electromagnetic radiation coming from the earth's surface. The images are faithful and instantaneous representations of portions of the earth's space seen from above, which in itself represents an original view. The vertical view was intellectually constructed long before technology could offer it, so we found cadastral plans engraved on clay tablets dating back to several centuries before our era and we can cite the painted vertical representations of cities or the representations of lordly domains. These intellectual constructions testify to man's quest to appropriate a space: to grasp it, to understand it, to develop it.

More specifically and briefly, remote sensing is the "Body of knowledge and techniques used to determine physical and biological characteristics of objects by measurements made at a distance, without physical contact with them." (Interdepartmental Commission on Aerospace Remote Sensing Terminology, ***1988.)*** This includes "the observation, analysis, interpretation, and management of the environment from measurements and images obtained using airborne, space, land, or sea-based platforms" (Bonn & Rochon, 1992).

1- History

The history of the evolution of remote sensing techniques can be divided into five major periods:

- The history of remote sensing began in 1858, when ***Gaspard Félix Tournachon***, known as ***Nadar*** (1820-1910), took the first aerial photograph from an aerostat over the Kremlin Bicêtre district in Paris.

- From the First World War to the end of the 1950s, aerial photography became an operational tool for cartography, oil research, and vegetation monitoring. There is a continuous progress of aviation, cameras and emulsions (color, black and white infrared, false color infrared). The methods of photo-interpretation are specified and codified.

- The period that begins in 1957 and ends in 1972 marks the beginning of space exploration and prepares the advent of current remote sensing. The launch of the first satellites, then of manned spacecraft carrying cameras, reveals the interest of remote sensing from space. At the same time, imaging radiometers were developed and perfected, as were the first on-board radars. The first operational application of space remote sensing appeared in the 1960s with the ESSA series of meteorological satellites.

- The launch in 1972 of the ERTS satellite (later renamed Landsat 1), the first remote sensing satellite for land resources, ushered in the era of modern remote sensing. The constant development of sensors and digital data processing methods opened up the field of remote sensing applications and made it an indispensable instrument for managing the planet and, increasingly, an economic tool.

- Since the 1970s, there has been a continuous development of remote sensing, marked in particular by :

- Increasing the spatial resolution of sensors.

- The diversification of sensors that use more and more varied and specialized areas of the electromagnetic spectrum. In the 1990s, we saw the multiplication of satellites equipped with active sensors, particularly radar. In the field of visible and infrared radiation, very high spectral resolution sensors are now commonly used in their airborne version and are making their appearance on board satellites.

- The dissemination of data on a commercial basis, envisaged since the launch of the SPOT program in 1986, is today reflected by the launch of remote sensing satellites by private companies. Remote sensing data are becoming the object of a competitive market. (Claude Kergomard, ENP-modified)

11- Tools involved in remote sensing

Remote sensing uses the properties of electromagnetic radiation to remotely analyze the surface of the land, ocean or atmosphere. A good knowledge of the elementary physics of radiation is essential for the interpretation of remote sensing results.

1. Electromagnetic radiation:

Electromagnetic radiation is a form of energy propagation in nature, the most familiar form of which is visible light as perceived by the human eye. Historically, physics specialized in the study of radiation (optics) was born from the study of the propagation of light and its interactions with materials (geometrical optics). Radiation was then recognized by physicists as a wave phenomenon, in relation to electricity and magnetism (electromagnetic optics). This perspective has considerably expanded the field of knowledge on the spectrum of electromagnetic radiation, well beyond visible light. Finally, modern physics has shown that electromagnetic radiation can also be considered as a displacement of elementary particles representing a quantity of energy (energetic and quantum optics).

1.1. Electromagnetic waves:

An electromagnetic wave is the simultaneous vibration in space of an electric field and a magnetic field. An electromagnetic wave is a progressive and transverse wave; the direction of the variation of the fields is perpendicular to the direction of propagation (figure 3).

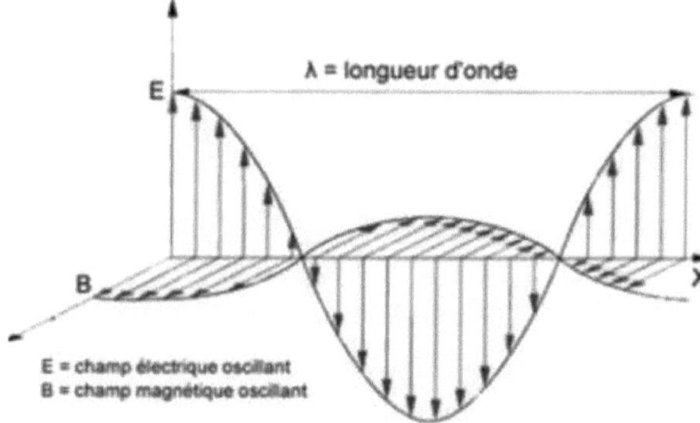

Figure 3. Electromagnetic waves.

The electromagnetic wave is characterized by :

> **the period T**: this is the time after which the electric or magnetic field regains its value from any time, ie makes a cycle. The unit is the second.

> **the frequency**, designated by the letter **r**: it is the number of cycles per unit of time.

The unit of frequency is the Hertz (Hz). One Hz is equivalent to one cycle per second. The waves used in remote sensing are characterized by very high frequencies measured in multiples of Hz (kHz, MHz or GHz -gigahertz).

> **the wavelength or amplitude λ**: it is expressed by a unit of length, the meter or its sub-multiples, in particular: the micron or micrometer (μm).

$1\mu m = 10^{-6}$ m and the nanometer: nm. $1nm = 10\ m^{-9}$

Between the wavelength and the frequency exists the classical relationship: $\mathbf{r * \lambda = c}$

Where is the speed of propagation of radiation in a vacuum (speed of light):

$c = 3*10^{8}\ m.s^{-1}$.

1.2. Radiation and energy :

The exchange of energy carried by electromagnetic radiation between the sun and the earth-ocean-atmosphere system is not continuous, but discrete, in the form of packets of energy, carried by immaterial elementary corpuscles, the photons. Each photon carries a quantum of energy proportional to the frequency of the electromagnetic wave considered; this energy is all the greater as the frequency is high.

The following relationship expresses the amount of energy associated with a photon as a function of the frequency of the wave: $E = h\nu$

where:

- E : the energy of the electromagnetic wave
- ν : the frequency of the wave
- h : Planck's constant (6,625.10-34 J.s).

1.3. The electromagnetic spectrum:

The electromagnetic spectrum represents the distribution of electromagnetic waves according to their wavelength, their frequency or their energy (Figure 4).

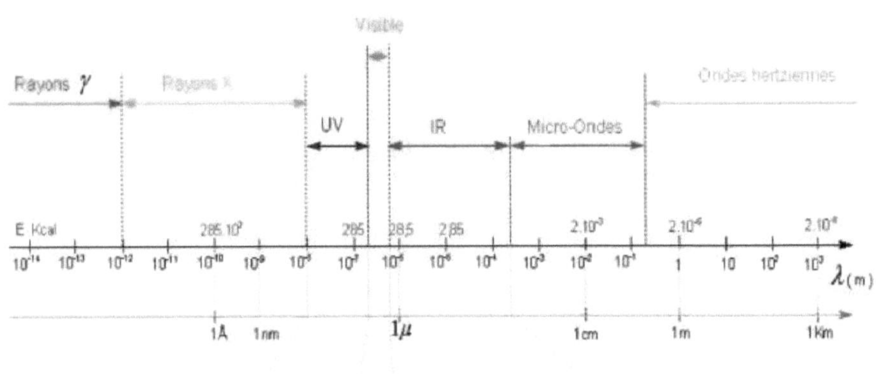

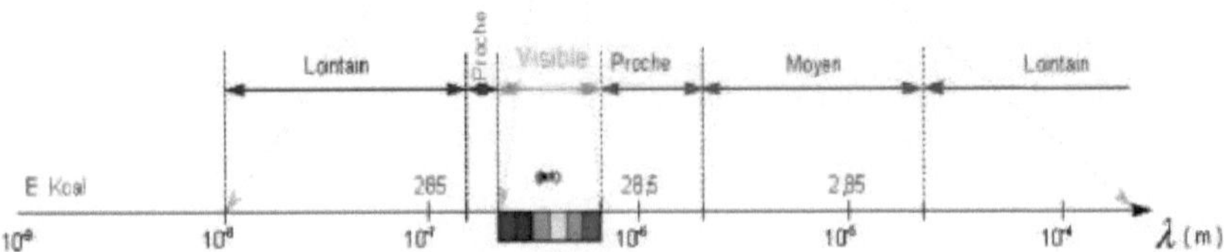

Figure 4. The electromagnetic spectrum.

Starting from the most energetic waves, we distinguish successively :

- **Gamma rays (γ):** They are due to radiation emitted by radioactive elements. Very energetic, they easily pass through matter and are very dangerous for living cells.Their wavelengths range from one hundredth of a billionth (10^{-14} m) to one billionth (10^{-12} m) of a millimeter.

- **X-rays :** Very energetic rays that pass more or less easily through material bodies and are slightly less harmful than gamma rays. They are used in medicine for X-rays, in industry (baggage inspection in air transport), and in research for the study of matter (synchrotron radiation).

X-rays have wavelengths between one billionth (10^{-12} m) and one hundred thousandth (10^{-8} m) of a millimeter.

- **Ultraviolet:** Radiation that remains quite energetic, it is harmful to the skin. Fortunately for us, a large part of the ultraviolet rays is stopped by the atmospheric ozone which acts as a protective shield for the cells. Their wavelengths range from one hundred thousandth (10^{-8} m) to four tenths of a thousandth (4.10^{-7} m) of a millimeter.

- **The visible domain:** Corresponds to the very narrow part of the electromagnetic spectrum perceptible by our eye. It is in the visible range that solar radiation reaches its maximum (0.5 μm) and it is also in this portion of the spectrum that we can distinguish all the colors of the rainbow, from blue to red. It extends from four tenths of a thousandth (4.10^{-7} m) - *blue* light - to eight tenths of a thousandth (8.10^{-7} m) of a millimeter - *red* light.

- **Infrared:** Radiation emitted by all bodies whose temperature is higher than absolute zero (-273°C).

In remote sensing, certain infrared spectral bands are used to measure the temperature of land and ocean surfaces, as well as clouds.

The infrared range covers wavelengths from eight tenths of a thousandth of a millimeter (8.10^{-7} m) to one millimeter (10^{-3} m).

- **Radar or microwave waves:** This region of the spectrum is used to measure the radiation emitted by the earth's surface and is similar in this case to remote sensing in the thermal infrared, but also by active sensors such as radar systems.

A radar sensor emits its own electromagnetic radiation and by analyzing the backscattered signal, it can locate and identify objects, and calculate their speed of movement if they are in motion. And this, whatever the cloud cover, day or night.

The microwave domain extends from wavelengths in the centimeter range to the meter range.

- **Radio waves:** This area of wavelengths is the largest of the electromagnetic spectrum and concerns the waves that have the lowest frequencies. It extends from wavelengths of a few cm to several km. Relatively easy to transmit and receive, radio waves are used for the transmission of information (radio, television and telephone). The FM band of radio sets corresponds to wavelengths of the order of a meter. Those used for cellular phones are about 10 cm.

Unlike the human eye, which is only able to capture radiation in a very narrow window of the electromagnetic spectrum, that corresponding to the visible range (wavelengths between 0),4μηι and 0),7μηι), satellite sensors use a much wider fraction of the spectrum.

Three spectral windows are mainly used in spatial remote sensing:

9. The field of the visible

10. The infrared domain (near IR, mid IR and thermal IR)

11. The field of microwaves or hyperfrequencies.

2. Radiation from the atmosphere:

When radiation passes through the atmosphere, it is partially or totally absorbed or scattered by the molecules that compose it. The radiation then gives up energy to the atmosphere.

We distinguish different types of scattering according to the relative size of the targets compared to the wavelength of the incident radiation. As we will see in the next part, solar radiation located in the Ultraviolet are absorbed in the upper atmosphere so that we consider mainly visible radiation:

■ Rayleigh scattering: is the scattering by molecules. Size of the target: 10 nm (nanometers, 10^{-9} m).

■ Mie scattering refers to the scattering by particles whose radius oscillates between 0.1 and 10 times the wavelength.

■ The scattering according to the geometrical optics (non-selective) occurs when the size of the target particles is very large compared to the wavelength.

3. The radiation of matter:

Subjected to a radiation emitted by an external source, the material absorbs a part of this radiation which is transformed into heat, the remainder is either reflected or transmitted through the body.

- **Emission**: depending on its temperature, a body emits more or less energy, it is distributed over wavelengths that also depend on its temperature. The earth's surface around 15°C (288 K) emits infrared radiation, the sun (6000 K) in the visible.

- **Transmission and absorption:** When radiation passes through the atmosphere, some of its energy is absorbed by the gases that make it up (nitrogen in the ultraviolet and infrared, water in the medium and thermal infrared, CO2 in the near infrared,...etc.).

- **Reflection:** the waves that reach a surface are reflected. The specular reflection concerns perfectly smooth surfaces. Most objects on the earth's surface are rough and the reflection is called diffuse. The fraction of the diffuse radiation returned to the transmitter is the backscattered radiation.

Reflectance: the proportion of light reflected for a given wavelength is called reflectance (Reflectance = Energy reflected / Energy received). (Russ, John C. (1995))

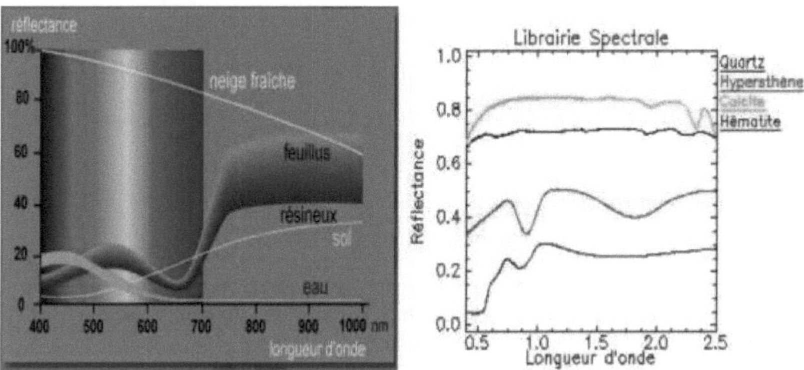

Figure 5: Reflectance of the different surfaces and their characterization by the different satellite channels.

4. Spectral signatures and water response:

Due to different reflective characteristics towards solar radiation, objects do not reflect the same amount of energy. The spectral signature of an object describes its reflective behavior in the entire electromagnetic spectrum. A priori, each type of object has its own spectral signature which is used in remote sensing to discriminate between objects. Figure 6 shows the spectral signatures of the most common objects on the Earth's surface. In particular, on this figure, water appears with a very different behavior from other types of objects. Indeed, the reflectivity of clear water is globally decreasing as the wavelength increases and tends to cancel out beyond the red spectral band (Bukata et al., 1995). This behavior is less pronounced for turbid waters, for which the spectral signature also incorporates the reflective characteristics of suspended particles.

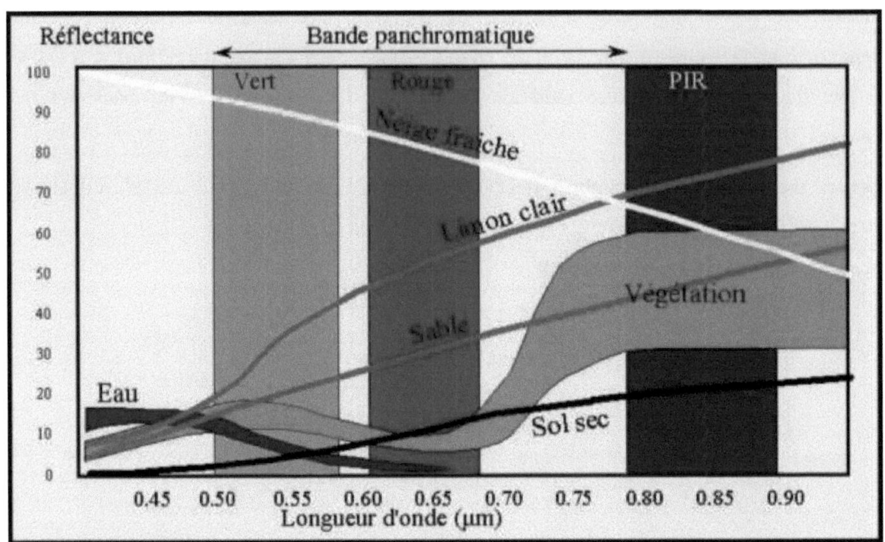

Figure 6. Schematic representation of the spectral signatures of the main objects on the Earth's surface - from (Maurel, 2001).

5. Vectors:

Depending on the distance to the ground, there are different types of vectors: those that operate at a few meters from the ground (cranes, or vehicles that support radiometers or cameras); those that operate between ten meters and ten kilometers (airplanes, helicopters, and balloons); those that operate between ten and a hundred kilometers (stratospheric balloons); and those that operate between 200 kilometers and 40,000 kilometers (the satellites)

12. Earth observation satellites:

Depending on the expected applications, Earth observation satellites occupy different orbits:

***J* Geostationary orbit**: A geostationary satellite is placed in an equatorial orbit (the angle between the orbital plane and the equatorial plane, or inclination, is zero) at an altitude of 35 800 km. It rotates at the same angular velocity as the Earth. It appears immobile for a terrestrial observer. It can only observe a part of the Earth. The satellites METEOSAT (France), GOES (Geostationary Operational Environmental Satellites, USA), GMS (Japan) and INSAT (India) are geostationary.

J Sun-synchronous orbit Satellites have a quasi-polar circular orbit (inclination - 90°) at an altitude between 700 and 900 km. The orbital plane is theoretically fixed, and the track (projection of the satellite trajectory on the Earth's surface) shifts by a certain angle towards the west due to the rotation of the Earth. These are scrolling satellites that can observe the entire surface of the Earth.

The French satellite SPOT and the American satellite LANDSAT circulate on this type of orbit.

J **Any circular orbit:** as for the TIROS, NOAA and ERS-1 satellites. We will limit ourselves to the description of the Landsat characteristics, because we have chosen the satellite images from the Landsat 7 ETM+ satellite, because they are a source of important geomorphological information. (Campbell, 1987)

> **LANDSAT image characteristics:**

The Earth Resources Technological Satellite (ERTS) program using the ERTS- 1 satellites, whose name has been changed to LANDSAT (Land Satellite), is due to NASA, with the aim of making multichannel shots of the earth's surface. All satellites in the LANDSAT series are sun-synchronous, in sub-polar orbit, with a standard altitude ranging from 917 (Landsat 1 to 3) to 705 km (Landsat 4 to 7), and pass over the same point every 16 days.

The first satellite, Landsat 1, was launched in 1972 and was followed by 4 others (Landsat 2 to 5). The first three are the first generation, equipped with two acquisition systems: the *RBV (Return Beam Vidicom)* digital camera and the *MSS (*Multi Spectral *Scanner*) multispectral sensor. In 1982, the Landsat 4 satellite is the first of the second generation, with a major change: a shift from a 4 to 7 channel acquisition system and a resolution of 30 m against 80 m before, and the last generation with Landsat 6, launched on October 5, 1993 and crushed at sea during the launch, and Landsat 7 successfully launched on April 15, 1999. Both equipped with new sensors: the *Thematic Mapper (TM) and the Enhanced Thematic Mapper Plus* (ETM+).

TM images are much more accurate than MSS because of their spatial, spectral, and radiometric resolution, but also because of the higher number of bands (Table 1). The Landsat 7 ETM+ instrument has 8 frequency bands (Lillesand et *al.,* 1994).

Tableau 1. Characteristics and applications of the TM sensor spectral bands.

Bands	Spectral range (pm)	Resolution	Application
TM 1	0.45 - 0.52 (blue)	30 m	Discrimination between soil and vegetation, bathymetry/coastal mapping; identification of cultural and urban features
TM 2	0.52 - 0.60 (green)	30 m	Green vegetation mapping (peak reflectance measurements); identification of cultural and urban features
TM 3	0.63 - 0.69 (red)	30 m	Discrimination between leafy and leafless plant species; (chlorophyll absorption); identification of cultural and urban traits

TM 4	0.76 - 0.90 (near IR)	30 m	Identification of vegetation and plant types; health and content of biological mass; delineation of water bodies; soil moisture
TM 5	1.55- 1.75 (short wavelength IR)	30 m	Sensitive to moisture in soil and plants; discrimination between snow and clouds
TM 6	10,4- 12,5 (thermal IR)	120 m	Discrimination of vegetation stress and soil moisture related to thermal radiation; thermal mapping
TM 7	2.08 - 2.35 (short wavelength IR)	30 m	Discrimination between minerals and rock types; sensitive to moisture content in vegetation

Tableau 2. Landsat 7 ETM+ band characters.

Frequency bands of the ETM+ instrument	Spectral bands	Spatial resolution	Wavelength
Band 1	Blue (visible)	30m	0.45-0.5 pm
Band 2	Green (visible)	30m	0.52-0.6 pm
Band 3	Red (visible)	30m	0.63-0.69 pm
Band 4	Near IR	30m	0.75-0.9 pm
Band 5	Average IR	30m	1.5-1.7 pm
Band 6/1	IR Thermal/far	60m	10.4-12.5 pm
6/2 band		120m	
Band?	Average IR	30m	2.08-2.35 pm
Band 8	Panchromatic (green-red-IR)	15m	520-900nm

6. Areas of application of remote sensing :

Remote sensing is applied to all disciplines that require the spatial distribution of a phenomenon, either to determine a state at a given time, or to follow a more or less rapid evolution of a phenomenon (Foin, 1985).

The first major field of application of remote sensing was the study of the atmosphere (meteorology and climatology). The interest of remote sensing in this field is to monitor the spatio-temporal evolution of cloud cover, measure temperature, water vapor and precipitation...

In oceanography and marine resources, remote sensing offers the advantage of allowing an analysis of ocean color (estimation of biological production, turbidity), and a study of the dynamics and characteristics of the seas and oceans (temperatures and surface altitude, waves and winds, turbidity of the coasts, ...), it also allows the monitoring of glaciers and icebergs

The terrestrial applications of remote sensing are extremely varied. They range from agriculture (crop yields, vegetation responses to certain environmental constraints, ...), forestry (forest mapping, estimation of certain dendrometric characteristics of forest stands, defoliation and health status, ...) and hydrology (spatialization of rainfall intensity, vegetation cover, ...), to urban planning and development (land use mapping, ...).), regular and thematic cartography, geology (recognition of the petrographic nature of surfaces devoid of vegetation cover, following the continental drift, thermal anomalies related to tectonic zones, ...), mining prospecting, geomorphology and structuring (identification of fault networks and thus the determination of preferential orientations of rupture) and natural hazards (drawing up "risk maps" for certain regions threatened by cyclones, earthquakes, volcanoes, land movements, drought, .).

111. Coupled contribution of remote sensing and GIS for hydrological study:

At the same time reference in space and witness in time, the basic geographical information constitutes an important contribution to the information on the environment. The techniques developed in the field of geographic information systems (GIS) and remote sensing are particularly relevant for studying a territory and its resources. They offer the advantage of considering a large number of different variables in the analysis, so as to address very diverse issues at different spatial scales.

We propose in this study to highlight the informational contribution of GIS and remote sensing for the prediction of flood risk areas.

1 ... Contribution of remote sensing to flood mapping :

Satellite imagery provides an objective view of fields with large spatial footprints and continuity over long streamlines (Ali and Qadir, 1989). These images are very useful and effective for determining flooded areas (Dhakal et *al.*, 2002; Giacomelli et *al.*, 1995). Works that exploit satellite imagery for mapping are relatively present in the scientific literature.

In studies based on optical image analysis, remotely sensed flooded areas are consistent with in situ observations (Rango & Anderson, 1974). However, significant errors in the estimation of flooded areas can occur when the resolution is too coarse compared to the size of the observed water surfaces (Imhoff et al. 1987), when water is masked by overburden (e.g. forest) over large areas, or in the presence of floating macrophytes that increase the radiometry of the water in the near infrared (Smith, 1997)

Flood images are also used to characterize the risk and impacts of floods (Moussa & Laranier,

2004).

a. Theoretical probability of obtaining usable satellite images of floods

In this context, (Joveniaux, 1986) has quantified the theoretical probability of obtaining several flood images that can be used for flood studies for several optical satellites. It appears from this study that the factors that influence this probability are the frequency of revisit of the satellite and the probability of the presence of cloud cover. The case of radar space images is a little different because the quality of the images is not affected by cloud cover. The probability of obtaining an image is therefore closer to the satellite revisit frequency. However, an element to take into account for radar sensors is the presence of wind that can make the water rough and difficult to detect on the images for certain polarizations and certain wavelengths.

b. Detection and mapping of floods by spatial remote sensing

Satellite images (radar or optical) of rivers in flood are rich in information that is very relevant in the context of operational flood management. However, this exploitation of satellite images of floods is restricted to a two-dimensional geometric characterization (contours and surface) of the flooded area without consideration for the hydraulic phenomena involved. Consequently, this exploitation of satellite images could be improved by a characterization of the flood in three dimensions and consistent with the physical phenomena and an integration of spatial characteristics.

For this we will use two stages of treatment:

First step: Extraction of spatial flood features from satellite images and analysis of their local relevance from a hydraulic point of view,

Second step: Cross-referencing relevant information from the images with a fine Digital Terrain Model (DTM)

Flood mapping is only a partial exploitation of satellite images. Indeed, as shown by some works (Horritt, 2000; Smith et *al.*, 1996), satellite images of floods are rich in information, beyond the flood maps, which can be very beneficial to flood management and in particular to hydraulic modeling. For example, a fine three-dimensional characterization, coupled with hydraulic modeling, would allow a more complete exploitation of satellite images.

2. Remote sensing and mapping of surface conditions :

The use of remote sensing for mapping surface conditions makes it easy to envisage a spatial extension of soil suitability characterization (Casenave and Valentin, 1989).

Based on field observations describing surface conditions, the proposed mapping procedure (Lamachère and Puech, 1995) consists in decoding satellite images to define four thematic planes:

- a map of the waterways (hydrographic network)
- a *vegetation* plan based on the density of the vegetation cover,
- a *soil* plan, differentiating the soils by their surface luminance,
- a *land use* plan, separating cultivated and uncultivated areas.

This decoding makes it possible to characterize each radiometric class, resulting from the digital processing of the images, by their composition in typical elementary surfaces. For each radiometric class, the passage of the four primary variables (water, soil, vegetation, land use) to the composition of elementary surfaces types from field observations.

The crossing, in a Geographic Information System, of the four thematic plans resulting from the decoding also allows the mapping of homogeneous hydrological units characterized, thanks to the field observations, by their composition in elementary surfaces.

3 **.image processing techniques :**

a. **Supervised Classification:**

This classification, still called Computer Assisted Photo-Interpretation (CAPI), is based on the visual interpretation of images by an operator. The latter delimits, under a Geographic Information System (Davis & Simonett, 1991), the objects of interest, for example a flooded area. This approach is particularly relevant for a qualitative preliminary analysis of images (Henry, 2004), or as a complement to other image processing methods (Yésou et *al.*, 2000). The main disadvantage of this method is that it leaves a lot of room for the operator's subjectivity.

b. **Unsupervised classification**

This classification, also known as pixel to pixel image classification, is based on a partition, in classes Ci, of the space of radiometries of the image, the dimension of this space being equal to the number of spectral bands of the starting image. To each class Ci we associate a characteristic value Vi. The classification of an image pixel to pixel is then to create a new image with the value Vi if the pixel considered has intensities in each band of the initial image belonging to the class Ci.

A particular case of image classification is radiometric thresholding on a spectral band. This consists in creating a binary image (0-1) by applying a threshold function to the values of the pixels of the image in the spectral band considered. This method is commonly used for mapping objects whose radiometry is very contrasted with that of other objects in the image, such as flooded areas (Henry, 2004; Horritt et al, 2003; Maurel, 1988; Puech, 1992; Zhou et al, 2000). It is efficient and relatively easy to implement. One of the main difficulties of this method is the determination of a threshold value in the case where no additional data concerning the flooded area is available (Ryu et al, 2002).

c. **Segmentation by active contour models ("Snake")**

Active contour models are two-dimensional curves with moving nodes. Image segmentation using these models is based on the computation of local image statistics and on the shape growth of an initial polygon (or polyline). This initial polygon is spatially expanded to match the contours of the object of interest. The parameters associated with these algorithms are relatively numerous and include in particular the noise tolerance and the curvature of the polygon lines. The dynamics of the polygon contours is based on the notion of internal and external energy, the goal being to minimize the total energy present along the curve. In particular (Horritt, 1999; Horritt et al., 2001) proposes an active contour model algorithm to extract the hearing from a flood radar image.

d. **Change detection**

For flood mapping purposes, change detection requires the availability of at least two images (with similar radiometric and spatial characteristics), one of which is not flooded and the other is flooded. In the case of optical images, it is necessary to have common spectral bands on both images. In the case of radar, it is also necessary that the acquisition conditions - viewing angle and direction of orbit - are similar (Gineste, 1998; Henry, 2004). Indeed, in the case of viewing angles that are too far apart or different orbit directions, it is very difficult to match the images in space because of geometric distortions (see § 2.2.5). For optical images, change detection can be based on a Principal Component Analysis (PCA) on the different spectral bands or a radiometric pixel-to-pixel difference (or ratio) of the images - or on a sliding window - (Badji & Dautrebande, 1997; Delmeire, 1997; Pelletier et al, 2005). For radar images, a radiometric difference - or ratio - or a speckle decorrelation measure (Rignot, 1993) can be used.

4 Use of satellite images to assist in hydraulic modeling

Hydraulic models are very relevant for flood management support (De Roo et *al*., 2000; Horritt & Bates, 2001; Pelletier et al., 2005). However, the determination of parameter values for these models is often complex and challenging. Satellite images of floods, with an objective spatial vision, can be very useful for hydraulic modeling, for example for the calibration, validation, or inversion of hydraulic models.

The calibration of a model consists in optimizing, by successive simulations, the parameters in order to obtain simulated results as close as possible to the observed data.

***J* Integration of spatial data in hydraulic modeling:**

The use of spatial Earth observation data to assist hydraulic modeling is relatively recent and under development. The work currently done uses mainly the flooded areas extracted from satellite images

Conclusion

In general, studies based on the analysis of satellite images acquired during flood periods provide a fine two-dimensional characterization of the hazard, but without consideration for water levels and flood dynamics.

Accordingly, the state of the art in this chapter leads to the following observation:

Satellite images (radar or optical) of rivers in flood are rich in information that is very relevant for flood management, but which is most of the time only partially exploited.

Indeed, the richness of information of flood images goes beyond a two-dimensional map of the hazard (Horritt, 2000; Pappenberger et al, 2005a; Smith et al, 1996). Furthermore, (Raclot, 2003a) estimated water levels by crossing information extracted from aerial flood photographs and a fine DTM. In order to obtain uncertainties adapted to river hydraulics, he completed the crossing of information by a hydraulic coherence algorithm. In view of these two approaches, it seems very interesting to couple them, in order to obtain estimates of water levels from satellite images of floods, with uncertainties adapted to hydraulic modeling

Part II. Presentation of the study area & Methodological approaches.

Chapter I: Presentation of the study area

I. Geographic and administrative setting:

Oued Medjerda, known as Bagrada in ancient times, originates in Algeria and crosses northern Tunisia in a SW-NE orientation to flow into the Gulf of Tunis, between Cape Farina (also known as Rass Ettarf or Cape Sidi Ali El Mekki) and Cape Gammart.

It has a length of 460 km and a watershed area of about 22,000 km^2 . The latter includes, in Tunisia, 5 main tributaries: Oued Beja, Oued Tessa, Oued Kessab, Oued Seliana and Oued Mellègue (Oueslati et *al.*, 2006).

The Medjerda River Basin is subdivided into three sub-basins, namely

- **The high valley**: It extends from the city Ghardimaou, on the Algerian border, to the dam Sidi Salem.
- **The middle valley**: It extends from Sidi Salem dam to Laaroussia dam.
- **The lower valley**: It extends from Laaroussia dam to Ghar El Melh

The study area is located in the middle valley of the Medjerda between the downstream of the Sidi Salem dam and the upstream of the Laâroussia dam. It is attached administratively to the delegations Oued Ezzarga, Medjez Elbab and Tebourba (Figure 7).

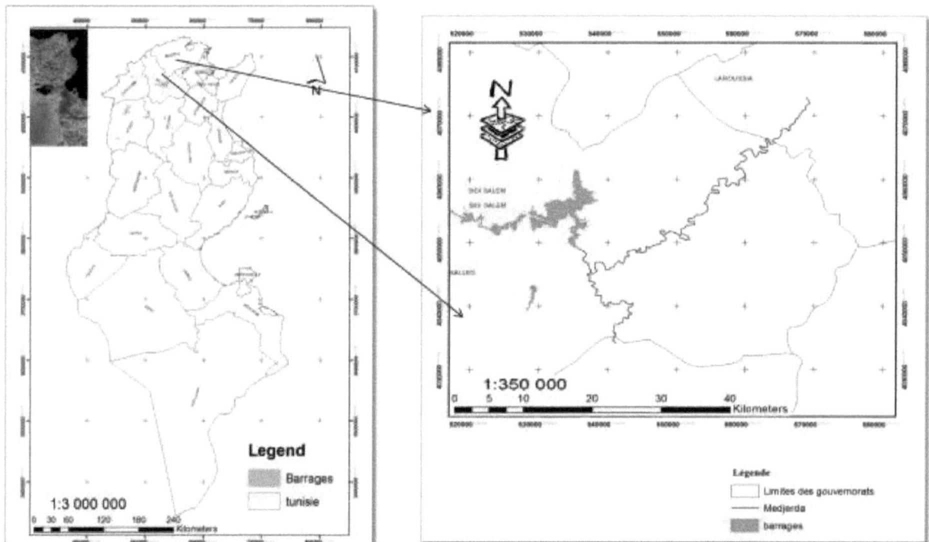

Figure 7. Location of the study area.

II. Climate setting:

Temperature, wind, evaporation and rainfall are the main climatic factors that influence the hydrological regime of the Medjerda watershed.

1. Rainfall:

The annual rainfall of the station Medjez Elbab show a great oscillation and show an annual average of about 445 mm.

However, the rainfall regime in the Medjerda is unstable and irregular since Tunisia is located on a climatological discontinuity where the slightest cause can have very important effects on its rainfall regime.

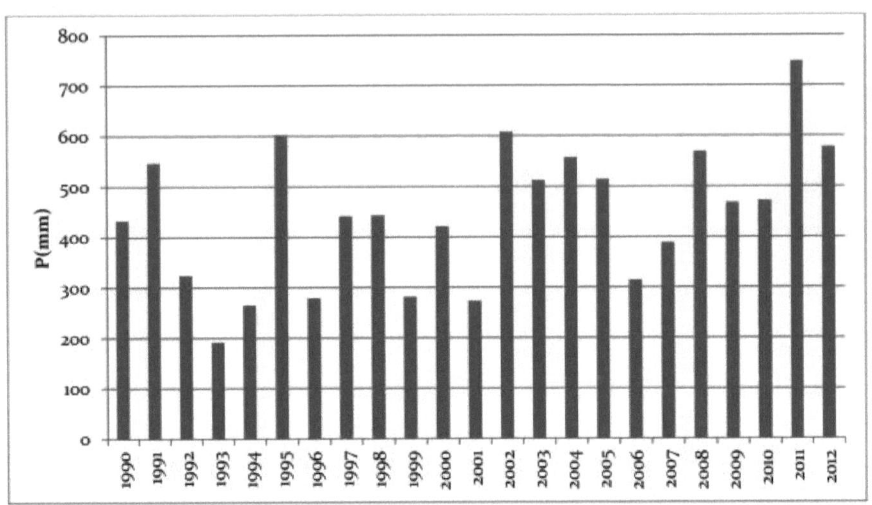

Figure 8. Variation of annual rainfall in Medjez Elbab station (DGRE, INM, 2014)

The study section, located between the Sidi Salem dam and the Lâroussia dam, is controlled by the rainfall stations presented in the following figure:

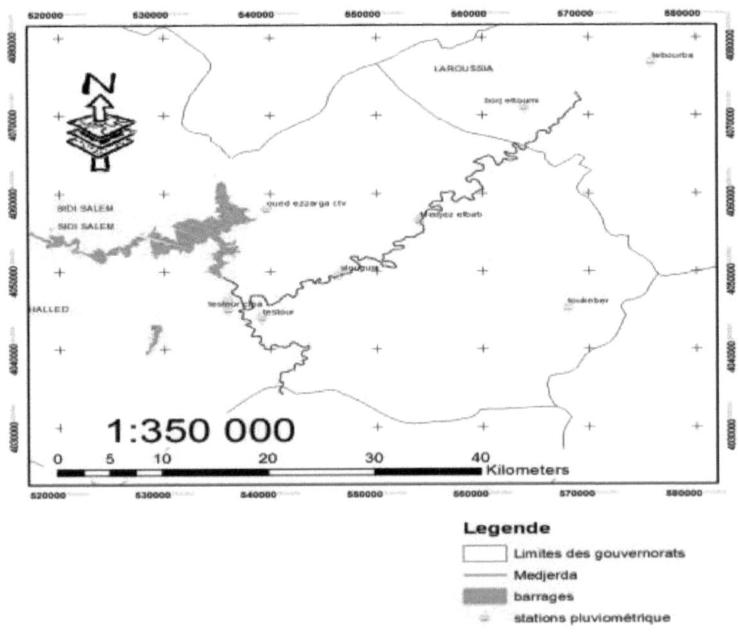

Figure 9. Rainfall stations in the middle Medjerda valley.

2. Temperature:

The temperature varies around 30°C between June and August. The minimum value recorded is 8°C to 14°C between January and March (figure 10).

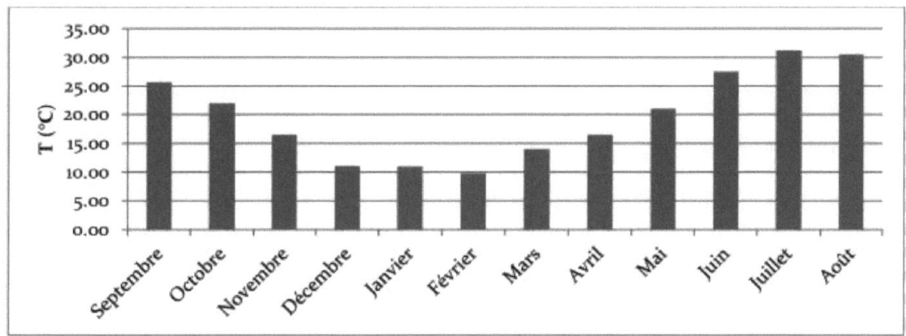

Figure 10. Variation in monthly average temperature in 2012 (DGRE, INM,2014).

3. Wind:

The prevailing wind is from the northwest. Strong to very strong winds are quite frequent in calm weather. A particular wind, the Sirocco, whose drying effects are considerable, blows from the South to the Southwest. This wind blows more than twenty days a year in the Medjerda basin.

4. Evaporation:

In 2009, the annual evaporation of Medjez Elbab was equal to 1176.1 mm.

The summer months (June, July and August) have the highest values of evaporation, reaching a total of 571.5 mm, i.e., half of the average annual value.

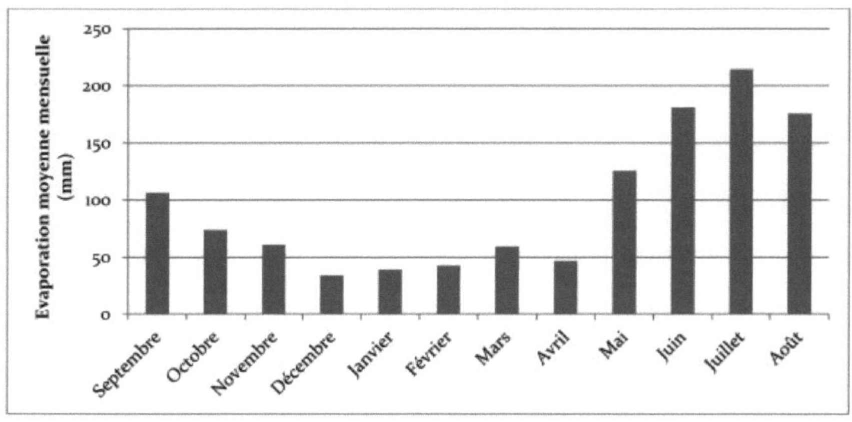

Figure 11.Monthly records of evaporation at the CRDA of Medjez Elbab.

5. Bioclimatic floor

The bioclimatic stage is calculated by the empirical formula of Emberger:

$$Q=\frac{2000\times P}{T^2\max - T^2 min}$$

With Q: A coefficient that allows to classify the region according to the bioclimatic.

P : Average rainfall (mm).

T_{max}: Average of the maxima of the warmest month (°K).

T_{min} : Average of the maxima of the warmest month (°K).

T_{max} = 31.2 + 273 = 304.2 °K

T_{min} = 9.8 + 273 = 282.8 °K

So Q = 70.82

⇒ According to Emberger's classification, M= 70.82 belongs to the interval]50, 70[, hence our area is part of the upper semi-arid. (Appendix 6).

III. Hydrological framework:

1. Surface water :

a. Oued Medjerda (section Sidi Salem - Lâaroussia)

b. Oued Khalled:

It originates in the plain of Krib, it drains a catchment area of 452 km at its confluence with the Medjerdah around Testour just upstream of the confluence of oued Siliana with the Medjerdah.

c. Oued Siliana

Oued Siliana is one of the main tributaries of the Medjerdah. It drains a catchment area of 2220 km2 at its confluence with the Medjerdah in Testour.

d. Oued Lahmar

It drains the plain of Goubellat, its catchment area covers a surface of about 520 km2.

2. Hydraulic works :

The section of Oued Medjerda subject of this study is limited by two dams:

a. Sidi Salem dam:

The Sidi Salem dam is considered the largest dam in Tunisia, it was built in 1981 as part of the master plan for the use of water in northern Tunisia and intended to mobilize the maximum resources of the large basin of the Medjerda.

In addition to the mobilization of surface water for the satisfaction of various water needs, Sidi Salem plays an essential role in the protection of cities and plains downstream (such as the city of Medjez Elbab,..) at the time of floods and this by laminating floodwater (Louati, 2011)

b. Lâaroussia dam:

It is a regulation work. It is located on the lower stream of the Medjerda at the entrance of the lower valley. It is located in the delegation of Tebourba, Governorate of Manouba.

It is, in fact, a mobile dam, consisting essentially of three large sector gates. It has as objective:

- To control the flows sent to the Medjerda-CapBon canal.
- Provide an annual volume for irrigation.
- To produce electricity by turbining.

c. Bridges:

The Sidi Salem-Lâaroussia reach contains five bridges:

- Testour.
- Slouguia.
- Andalusian (Almouradi) in Medjez Elbab.
- GP5 belt.
- Borj Ettoumi.

IV. Geological setting:

1. Lithostratigraphy:

The geological outcrops of the study area are essentially constituted by Pliocene formations of continental origin composed by conglomerates, sand and sandstone. All these sediments fill a depression limited on both sides by Cretaceous and Eocene formations where diapiric spikes of Triassic water appear and eventually seep into the water table (Clanzig et *al.*, 2010).

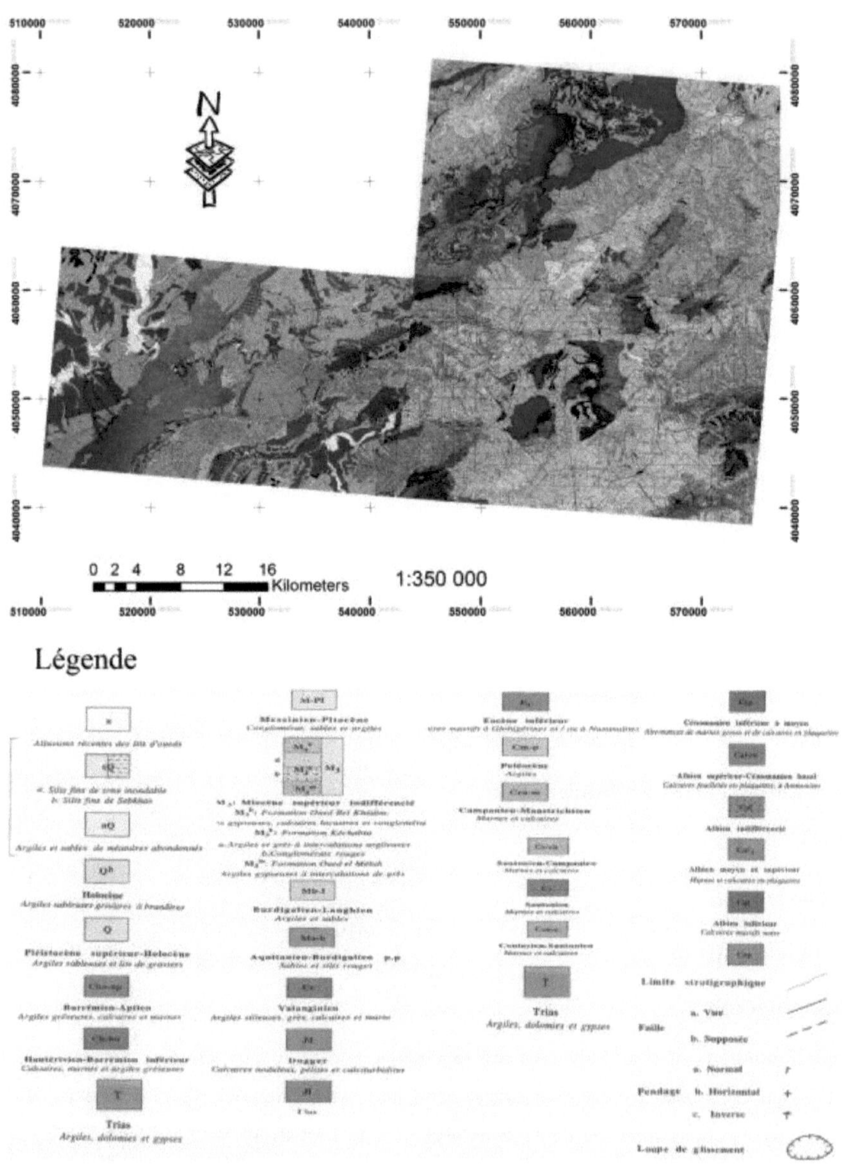

Figure 12: Geological map of the middle Medjerda valley.

2. Geomorphology:

From the Digital Terrain Model presented in the following figure, we notice that the relief of the middle valley is quite strong. The altitude in this area varies between 33m and 720m. It presents an alluvial plain which widens from the Medjez ELbab.

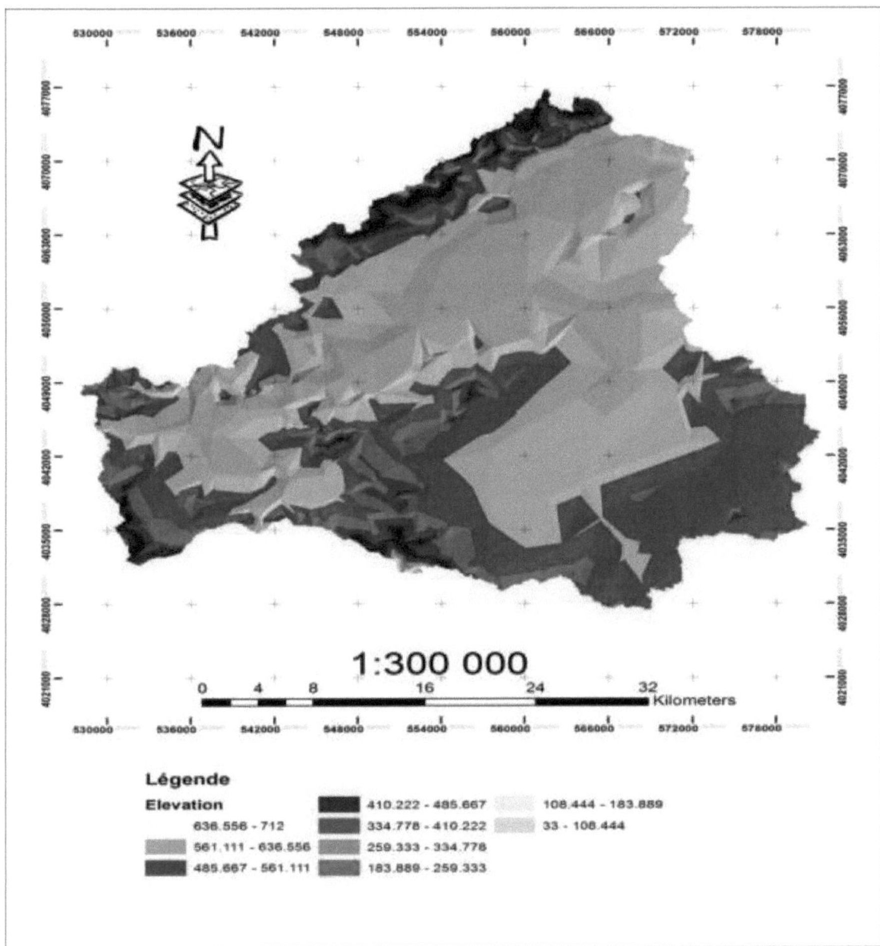

Figure 13.Digital Terrain Model of the middle Medjerda valley.

3. Structure:

From Testour to Tebourba, the middle valley of Medjerda is a gutter installed in the axis of the diapirs zone. These, arranged on either side at Slouguia. Moreover, it runs along the front of the T4 accident. These two features suggest the structural complexity of the region and the existence of transverse accidents caused by offsets and torsions that affect the geological basement.

The Testour depression is located at the intersection of the T4 accident and the Oued Ezzarga-Testour transverse tear. Up to the Lâaroussia dam, the Medjerda now flows in a horizontal staircase that it crosses in a single meander train.

Only the peculiarities of its design translate the passages from one step to another. The important transversal accident of the Oued has generated the unevenness of Medjez Elbab (Claude J et *al.*, 1981).

4. Pedology:

Although the soils of the mid-valley plain have a common origin, differences exist. Brown calcareous soils with textures that vary between silty, clayey and clayey. These soils are generally not very permeable and pose alkalinization problems (Rejeb et al, 2003). The texture of these soils favors the solid transport in suspension due to the dominance of particles of small diameter (clay, silt and sand).

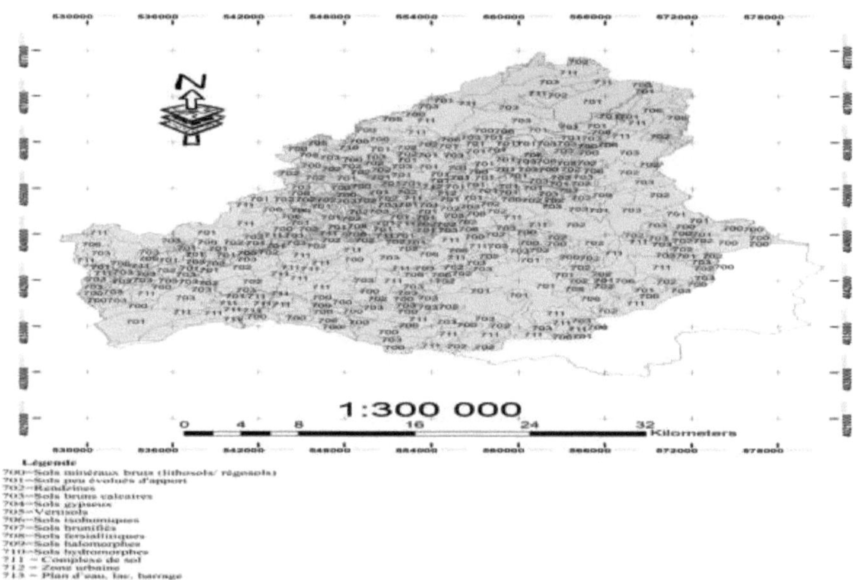

Figure 14: The pedological map of the middle Medjerd valley

V. Hydrogeological setting:

Almost all the structures containing aquifers are located in the plains (some are collapsed) or along the wadis. It is also necessary to point out the existence of some calcareous structures most often perched of rather limited extension" and whose resources are rather weak (Hechmet, 1989)

In this paragraph we will describe the different hydrogeological structures:

1. Water table

a. The water table of the plain of Teboursouk:

It is located at the foot of the Eocene limestone table of the city of Teboursouk, it extends on both sides of the wadi Khalled. The surface occupied by the table is quite limited it is about 47 km2.

The supply of the water table on the western side is ensured by the relief of the watershed of the Zitouna wadi whose quality of runoff water is good. On the eastern side it is the triassic massif of Jebel Ech Cheid which feeds the water table. Unfortunately, the runoff from this area contributes to the degradation of the chemical quality of groundwater.

b. The water table of Bled Ghenimah (Testour):

The water table of Bled Ghenimah develops in the lower course of the wadi Khalled just before its confluence with the Medjerda and develops along the Medjerda to the vicinity of the village of Sloughia. It should be noted that the water table of Bled Ghenimah covers an area of 30 km.

Almost all the potential of the water table comes from the supply of the latter from the releases of the Sidi Salem dam.

The dry residue of the whole table varies from 1.5 g/1 to over 7 g/1. The area where the salinity is between 1.5 and 3 g/1 is located at the level of the Khalled wadi and its confluence with the Medjerda.

c. The water table of the Goubellat plain:

The plain of Goubellat is located 15 km south of the village of Medjez el Bab. It is limited NE by the Jebel Bassina, to the West by the Jebel Djaffa, to the SE by the hills of Sidi Ramdane and Mahsoudj.

The water table is fed by the landforms that delimit it. In the zone where the depth of the water body is close to the natural ground, the salinity is very high, above 7 g/l. In the rest of the water table, the salinity is between 2 and 5 g/1.

2. Deep water :

a. The deep water table of Bled Ghénimah:

The deep aquifer of Bled Ghénimah is only recognized at the confluence of the Medjerdah, Khalled and Siliana wadis. The aquifer is composed by an alternation of gravel, pebbles and clays.

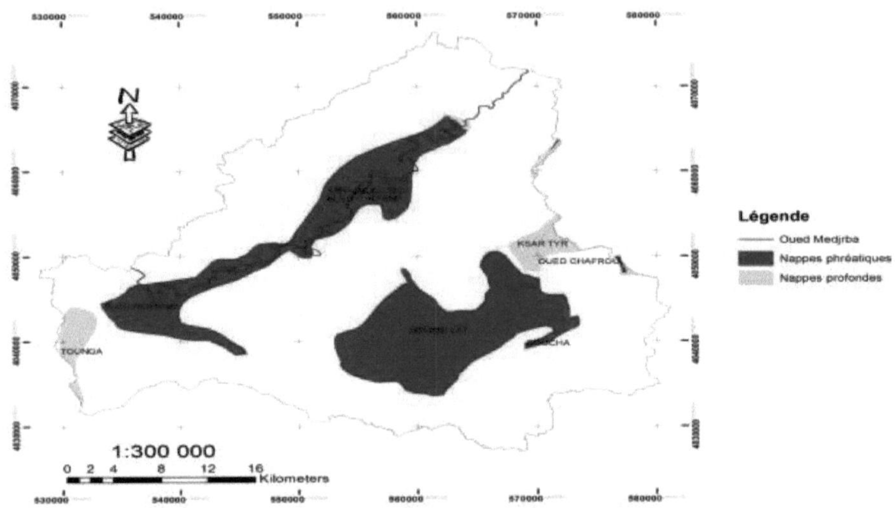

Figure 15. Map of the location of groundwater and deep water (STUDI, 2008)

Conclusion

In this chapter, we have presented the middle valley of the Medjerda by determining the geographical and administrative framework, the climate of the area, its hydrological framework. We have also shown the geology by detailing its lithostratigraphy, its structure, its pedology and its relief. On the same axis, we have presented in this chapter the different types of existing aquifers.

Chapter II: Methodological approach

Introduction

In this chapter we propose to present a methodological approach based on the choice of several data sources (satellite images, Google Earth extracts, topographic maps, geological maps... ...at different dates and different scales) for the mapping of flood risk areas.

Among the methodological problems raised, in this work, is the harmonization of heterogeneous data according to the same geographical reference frame (UTM projection, Datum Carthage, zone 32 N).

I. Remote sensing

The methodological approach adopted in this work is summarized in Figure 16.

1. Satellite images used

The present study was based on the processing and interpretation of ***Landsat*** satellite images for different dates, covering the region of the middle Medjerda valley: ***Path 192, Row 034, and Path 192, Row* 035**.

- A ***Landsat 7 ETM+ multispectral and panchromatic*** scene [one TM 8 panchromatic band at 15 m spatial resolution, [6 multispectral bands (TM 1, 2, 3, 4, 5, 7, at 30 m)], dated ***January 04, 2003***;

- Two ***Landsat 5 TM multispectral*** scenes [6 multispectral bands (TM 1, 2, 3, 4, 5, 7, at 30 m)] dated ***November 28, 2009*** and ***April 21, 2010***;

- A ***Landsat 8 OLI_TIRS multispectral and panchromatic*** scene [one TM 8 panchromatic band at 15 m spatial resolution, 6 multispectral bands (TM 1, 2, 3, 4, 5, 7, at 30 m)], dated ***25 December 2013***.

The selection of these images is made according to the following criteria:

S The year of acquisition (pre- and post-flood).

S The season of acquisition (ideally the same season).

S The time of acquisition (as much as possible the same).

2. Pre-treatments

Different pre-processings were performed for the available satellite images.

a. Geometric correction:

The quality of geometric corrections of images is essential in remote sensing of changes. These corrections allow the comparison of scenes positioned in the same geographic reference frame (UTM), as well as their superposition on the ortho-photo database or other data sources.

b. Atmospheric correction:

Atmospheric corrections are performed on satellite images of the study area. This type of correction is essential and has the following objectives:

J To access physical values of surface reflectance;

J Perform multi-date comparisons between images from the same sensor or from different satellites;

S To ensure the reproducibility of surface identification or classification methods, without having to repeat the analysis of samples taken from the image to be processed (Kergomard, 1990) (Appendix 1).

c. Stretching:

This technique facilitates the interpretation of satellite images, because it improves their visual quality. Indeed, it consists of a linear or non-linear transformation of the amplitude of the signal of each pixel of an image so that the set of amplitudes occupies more efficiently the available gray scale (RABIA M.C., 1998). (Appendix 2)

d. Field data acquisition :

The acquisition of field data is necessary to perform a supervised classification of remote sensing images and to refine our understanding of land use patterns. A field campaign took place in March in order to know the morphology of the land, its geology and especially the knowledge of land use and cultivation practices.

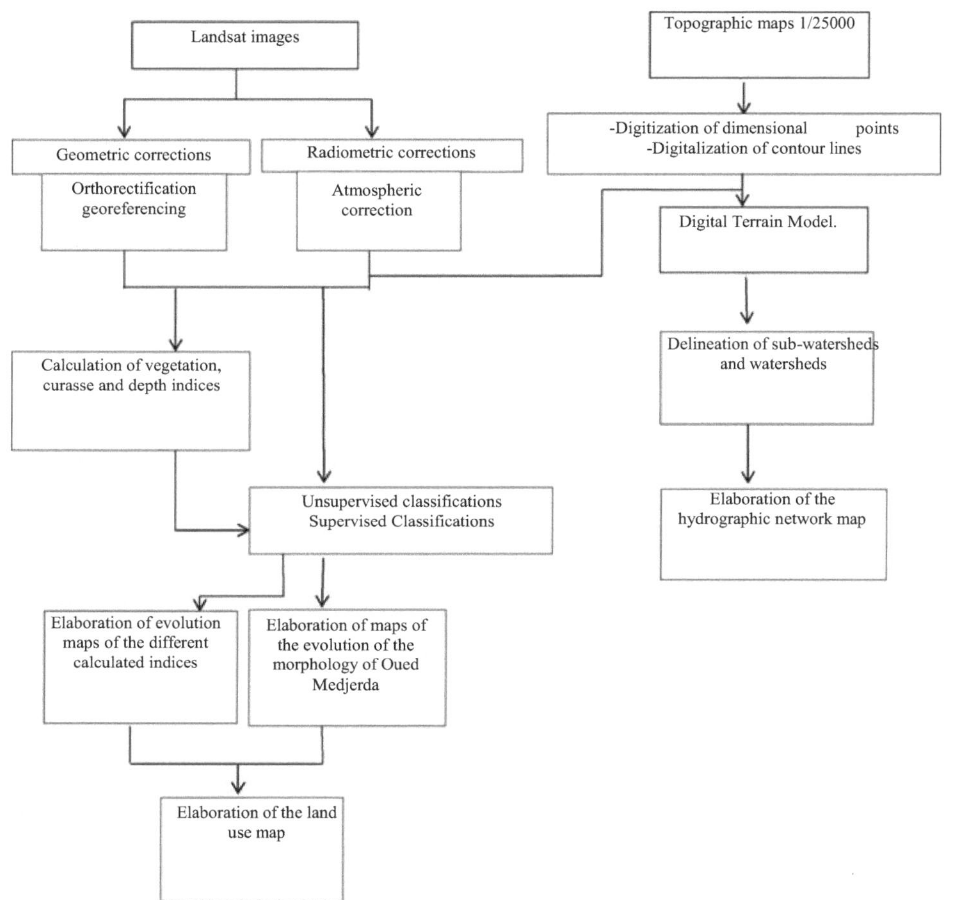

Figure 16. Summary diagram of the methodology adopted for the processing of satellite images.

e. Image classification and comparison :

***J* The principle**: The technique of comparing image classifications makes it possible to locate and identify changes in land use. A method has been applied to each image in order to compare them later. The difficulty is to reproduce exactly the same classification on each image. The supervised classification method used for this study is "maximum likelihood". The knowledge of the study area is capitalized and used to determine the membership of each pixel to a class. The supervised classification was carried out on georeferenced images in order to superimpose the layers of information used to digitize the classification kernels.

J The choice of classes: twelve classes were defined for the supervised classification of the two satellite images. This large number of classes allowed us to limit certain confusions. In particular, several types of forests, water surfaces and soils were discriminated. After synthesis, the final textural classification is composed of nine thematic groups.

J Quality assessment of supervised classifications: The evaluation of the performance of supervised classifications on Landsat 2003, 2009 and 2011 images was performed visually by comparing the results with other documents, such as google earth images and/or topographic maps.

J Masks and change detection: The detection of changes for a thematic class between the two dates (2003 and 2009) is possible through the use of masks.

3. Software used

- . **ENVI** (The Environment For Visualizing Images, developed by the company "ITTVIS").

It is a complete commercial software for viewing and processing remote sensing images. It presents a logical and intuitive interface to read, visualize and analyze different image formats. All the image processing methods for geometric and radiometric corrections, classification and cartographic layout are present. In this work we used the ENVI 5.1 version.

- **Global mapper 15** is more than just a visualization tool capable of displaying the most popular raster images, elevation data and vector data. It allows to use GIS functionalities on the different datasets in a cheap and simple way. It also allows the visualization of elevation data in true 3D with a draping of any raster image or vector data.

In this work the Global mapper software was used in the generation of the raster DTM.

II. Implementation of a GIS

1. Topographic maps

Topographic maps at scale 1:25,000 have been used in this work. These are sheets No. 18, 19, 26, 27, 28, 34 of the regions of Beja, Tebourba, Oued Ezzarga, Medjez El Bab, Bir Mcherga and Bou Arada established by the Office of Topography and Cartography (Tunis, 1988). These maps were georeferenced according to the UTM projection (Zone 32, Ellipsoid of Clarke 1880), assembled and vectorized using ArcGIS 10 software.

These documents constitute a corpus of information layers that are of capital interest for our study. In order to build a multi-theme database, thematic maps have been elaborated such as :

- Watershed delineation
- Map of the hydrographic network
- The map of hydraulic structures (dams, lakes...)
- Map of existing facilities

2. Geological maps

Geological maps at the scale 1/50 000th established by Fournet (1999) of sheets 19, 26 and 27 of the regions of Tebourba, Oued Ezzarga and Medjez El Bab were thus georeferenced according to the UTM projection (Zone 32, Ellipsoid of Clarke 1880). These maps are assembled and vectorized in order to vectorize the geological map of the study area.

3. Software used: ArcGIS 10

The ***ARC GIS 10*** software was used in the georeferencing, the assembly of topographic and geological maps and in the establishment of a GIS of the middle valley of the Medjerda.

It is a system capable of acquiring, organizing, processing, analyzing and disseminating geospatial data. Designating both the software aspects and the methodological aspects. This software is a complete platform that exploits the geographical dimension in order to produce quality maps ready for use. It allowed us to exploit the results of classifications previously established from ENVI and present them in a clearer and more readable way in the form of maps.

III. Hydraulic modelling

The main objective of this study is the mapping of flood risk areas of the middle Medjerda valley which would be a decision support tool for planning and management of rivers and floodplains. This map is the product of the map of the flood hazard and the map of issues. The crossing of the simulation map with the land use map to extract, under GIS environment, several information allowing to anticipate the danger and then to protect the people and to minimize the damages.

1. Flood hazard map

It was based primarily on the use of hydraulic modeling of graded and free-surface flows under the 1D Saint-Venant model known as ***HEC-RAS*** *(Hydrologic Engineering Center - River Analysis System*) which is developed by the *Hydrologic Engineering Center* of the *U.S. Army Corps of Engineers* for flood simulation. Another main tool used is the ***HEC-GeoRAS*** extension through ESRI (Environmental Systems Research Institute) Arc GIS software either for preparation of data to be exported to ***HEC-RAS*** or to exploit the simulation results in a GIS environment (Debiane et *al.*, 2010).

2. Data used

The main data needed to carry out this work are the Digital Terrain Model (DTM) and the Manning's coefficient which represents the friction and which is deduced from the land use map, developed in the interpretation of satellite images, according to the nature of the type of soil (built, bare land, vegetation ... etc.).

3. Methodology of hydraulic flood modeling:

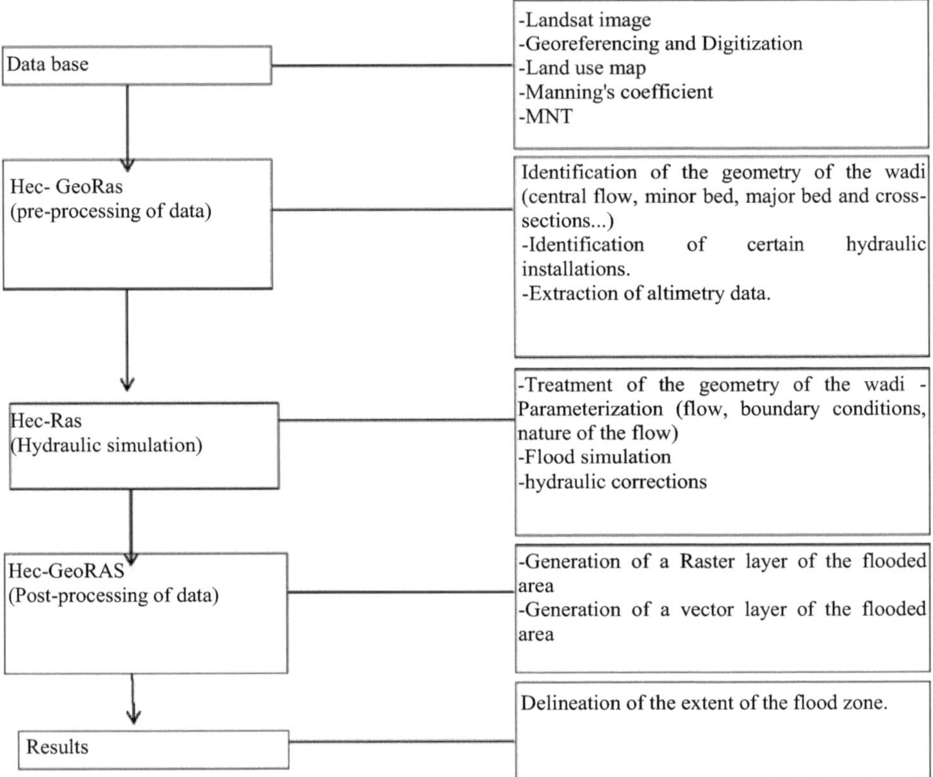

Figure 17. Summary diagram of the methodology adopted for the hydraulic modeling.

Conclusion

The methodology explained in this chapter divides the study into two main parts. The first part consists in extracting the maximum possible data from the acquired satellite images, and the second part consists in integrating the results of the hydraulic modeling with HEC-RAS in the GIS tool in order to map the flood hazard. The combination of the results from these two parts will lead us to the determination of the flood risk areas

Part III. Interpretations & Results

Chapter I : Study of the watershed of the middle Medjerda valley

I. Digital Terrain Model:

From the satellite image, we were able to establish the DTM of the study area (Figure 18)

Indeed, the Digital Terrain Model presented below, shows the fairly homogeneous relief of the middle valley of the Medjerda. Indeed, the altitude of the area varies between 0m and 710m.

It also indicates the remarkable surface of the alluvial plain of the area which is uniform from the downstream of the dam Sidi Salem to Medjez Elbab or it widens until the upstream of the mobile dam Lâaroussia.

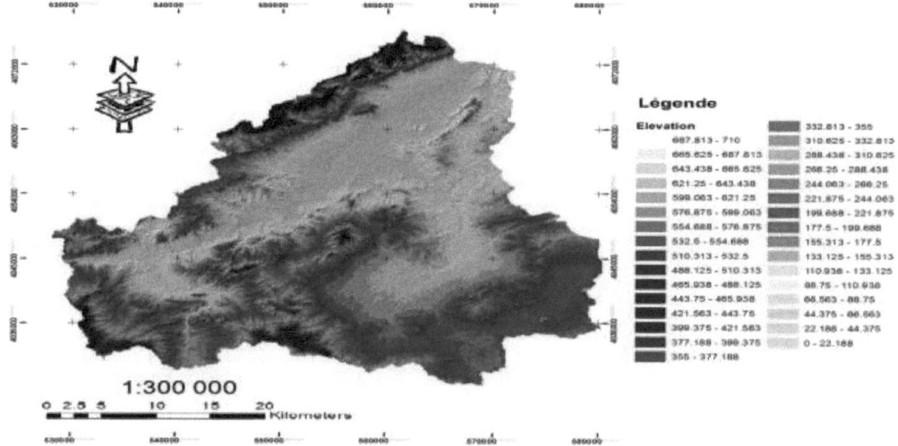

Figure 18. Digital Terrain Model of the study area.

II. Watershed and drainage system:

Using the ArcHYDRO extension of the ArcGIS environment, a hydromorphological analysis of the digital terrain elevation model was performed. This analysis can be summarized in five steps:

- Filling of bowls
- Flow management
- Accumulation of flows
- Flow vectorization

- Elaboration of watersheds

O *The results of the treatments are shown in the following figures.*

Tableau 3. The hydrological characteristics of the watershed of the middle Medjerda valley.

		Formula		***Unit***
Surface	***A***		**146800**	Ha
			1468	Km^2
Perimeter	***P***		**194**	Km
Length of the Thalweg	***lt***		**88**	Km
distance from the center of gravity to the outlet	***lg***		**19**	Km
Graveliu compactness coefficient	***kp***	-	**1.42**	without
The shape coefficient	***d***	-	**0.239**	without
Equivalent length	***L***	- (-)	**78.23**	Km
Equivalent width	***l***	- (-)	**18.76**	Km
	H5%		**0.395**	Km
	H95%		**0.047**	Km
the difference in altitude	***D***	H5%-H95	**0.348**	
Median altitude	***H50%***		**0.161**	Km
Average slope	***i***	-	**0.0001**	
Overall slope index	***Ig***	-	**0.004**	
Coefficient Of Specific Gradient	***Ds***	$I_g \times \sqrt{A}$	**0.17**	

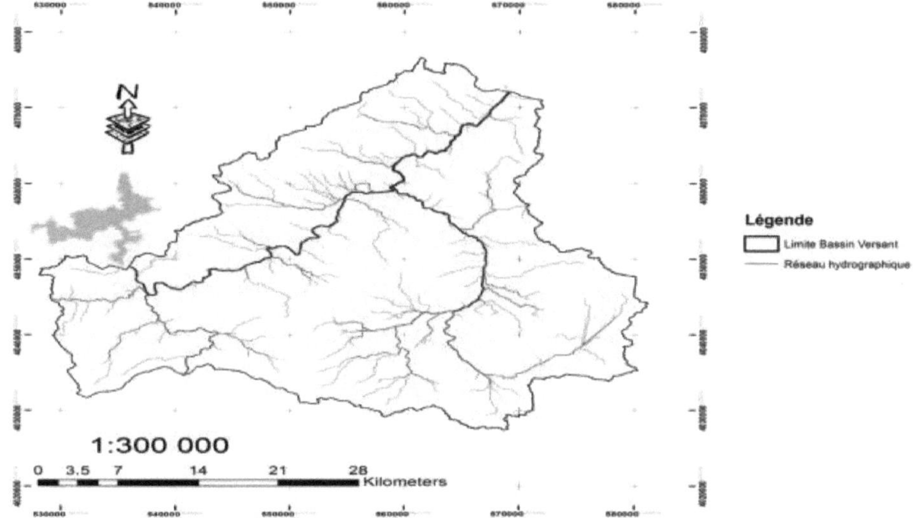

Figure 19. The watershed of the middle Medjerda valley and its hydrographic network.

From the calculations made earlier, it could be concluded that the shape of the watershed is fairly elongated to elongated and the relief is low. (Appendix 3, 4 and 5)

In addition, we note that the study area is fed by a varied hydrographic network. Apart from the water drained by the tributaries of Khalled, Siliana and Lahmar, the Medjerda wadi is fed by runoff from its watershed during torrential and irregular rains.

Indeed, the hydrological regime of the tributaries of the left bank of the basin of Oued Medjerda is more or less regular contrary to that of the tributaries of the right bank, where is located the section Sidi Salem - Lâaroussia.

111. The sub-watersheds :

The following figure represents the different sub-watersheds of the middle Medjerda valley. Indeed, Kalled, Siliana and Lehmar are the most important sub-watersheds of the region.

Tableau 4. The hydrological characteristics of the watersheds of the different tributaries of the middle Medjerda valley.

		KhalledSilianaLahmar					
			Unit		***Unit***		***Unit***
Surface	***A***	**45200**	Ha	**37500**	Ha	**52000**	Ha

		452	Km^2	375	Km2	520	Km2
Perimeter	*P*	105	Km	110	Km	97	Km
Graveliu compactness coefficient	*kp*	1.3927372	without	1.6018659	without	1.1995523	without
The shape coefficient	*d*	3.8372982	without	0.1695751	without	0.4923077	without
Equivalent length	*L*	41.646834	Km	47.025624	Km	32.5	Km
Equivalent width	*l*	10.853166	Km	7.9743758	Km	16	Km
	H5%	0.69	Km	0.66	Km	0.51	Km
	H95%	0.25	Km	0.35	Km	0.28	Km
the difference in altitude	*D*	0.44		0.31		0.23	
Median altitude	*H50%*	0.405	Km	0.505	Km	0.1825	Km
Overall slope index	*Ig*	0.010565		0.0065922		0.0070769	
Coefficient Of Specific Gradient	*Ds*	0.2246156		0.1276564		0.1613787	

The watersheds of the tributaries of the Medjerda in the middle valley have different forms:

- The basin of Oued Khalled: Quite elongated to elongated.
- The basin of Oued Siliana: Long to very long.
- The Oued Lahmar basin: Circular to fairly elongated.

The Medjerda watershed tends to lengthen from upstream to downstream and this is reflected by an increase in the compactness indices from upstream to downstream stations.

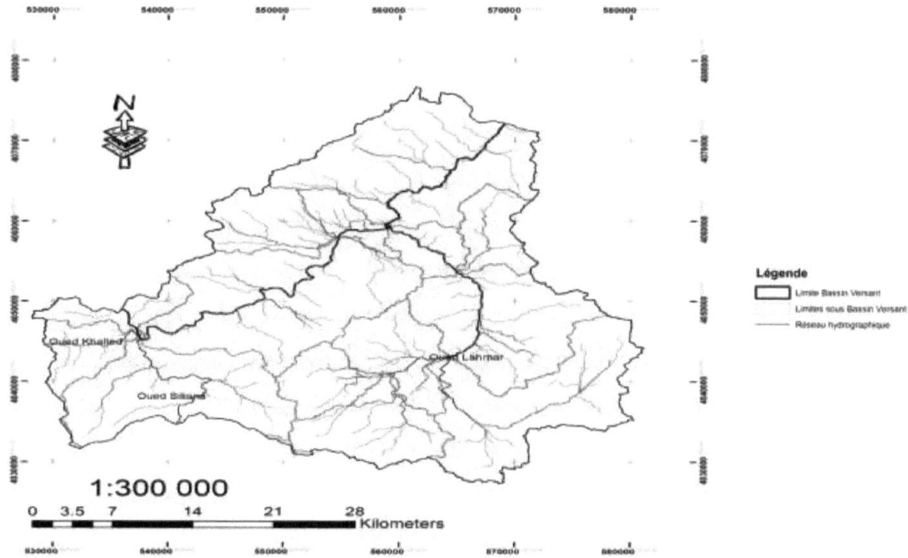

Figure 20. Delimitation of the Middle Medjerda Valley Basin and the Sub-basins Towards

Chapter II: Evolution of the morphology and land use of the middle valley of the Medjerda

Introduction

The analysis of some Landsat archive images allowed us to detect the different evolutions of the morphology of the Medjerda middle valley watershed.

The availability since 1972 of Landsat data, the oldest Earth observation program, makes it an exceptional documentary source. Indeed, Landsat data provide a global coverage of the Earth since 1972 thanks to the MSS (Multi Spectral Scanner, 1972-1992), TM (Thematic Mapper, since 1982 and still operational in 2006 with Landsat 5), and ETM+ (Enhanced Thematic Mapper, since 1999 and operational in 2000 with Landsat 7) sensors.

1. Evolution of the morphology of the Oued Medjerda section:

As already mentioned in the bibliographic part (chapter III) it is almost impossible to have optical satellite images during floods. Therefore, to study the evolution of Wadi Medjerda between Sidi Salem dam and Laârousia, we chose to interpret Landsat satellite images for three inter-flood dates.

- January 4, 2003: just before the flood of February 2003.
- November 28, 2009: After the 2003 and 2008 floods.
- December 25, 2013: After the November 2011 flood -

Each image was processed and interpreted, and the path of the wadi was digitized in order to compare its evolution for the three dates.

1. Landsat 2003 image

A Landsat 7 ETM+ image, dated January 4, 2003 in band composition 5,4 and 3, was processed and classified. From this image we digitized the route and morphology of the main river Medjerda for the year 2003 (Figure 21).

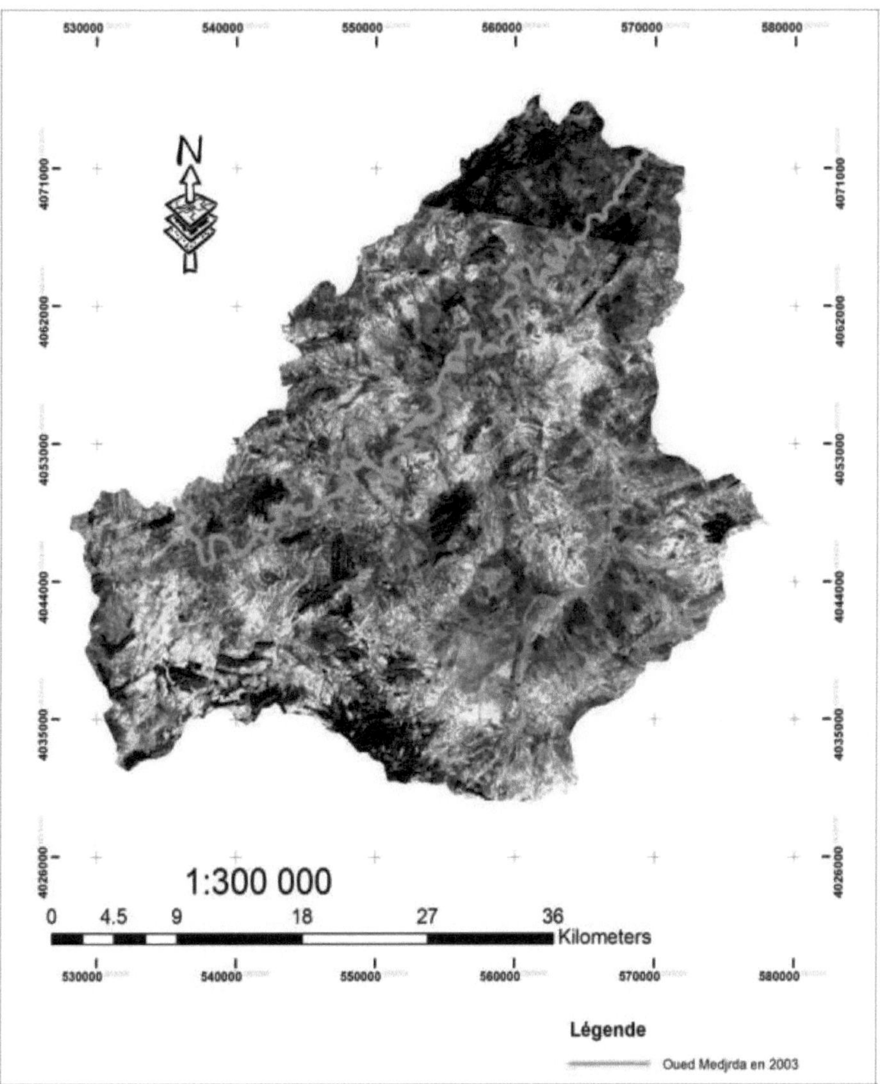

Figure 21. Processed and classified Landsat image (2003) showing the path of the Medjerda River (Sidi Salem-Laaroussia).

2. Landsat 2009 image

A Landsat 5 TM image, dated November 28, 2009 in composition of bands 4,5 and 6, was processed and classified. From this image we digitized the route and morphology of the main stream Medjerda for the year 2009 (Figure 22).

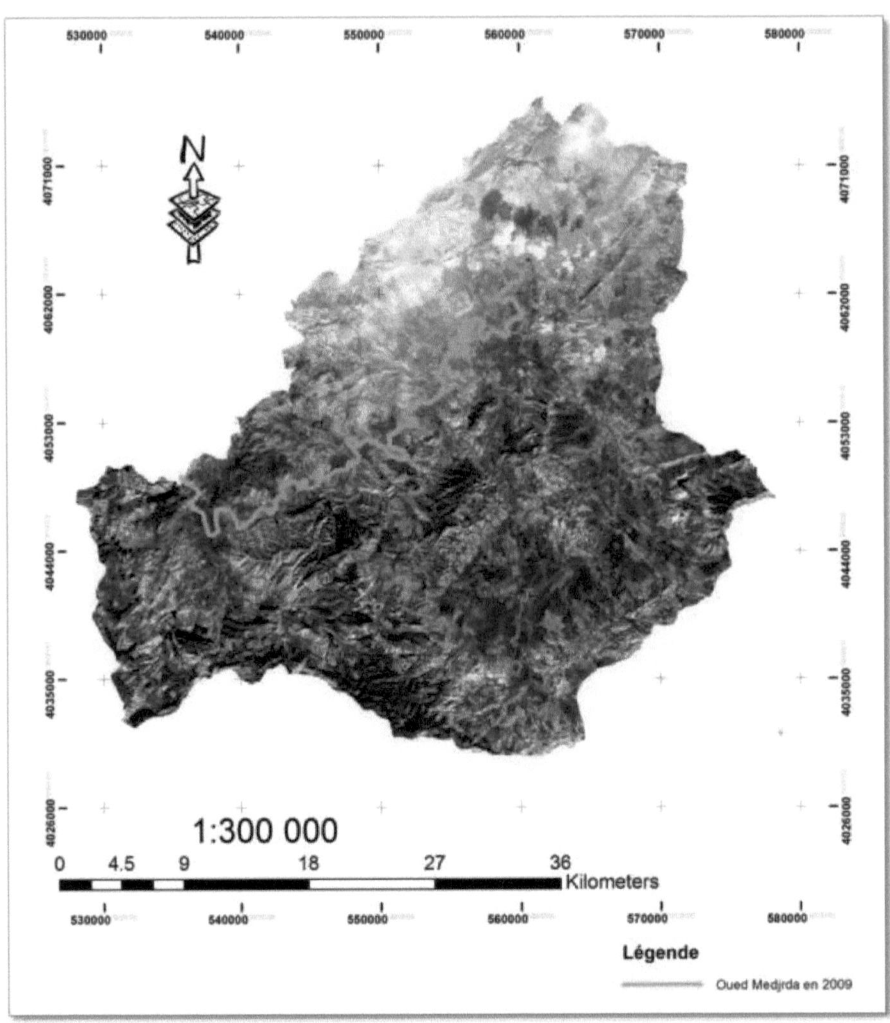

Figure 22: Processed and classified Landsat image (2009) showing the path of the Medjerda river (Sidi Salem-Laaroussia).

3. Landsat 2013 image

A Landsat 8 OLI_TIRS image, dated December 25, 2013 in band composition 6,4 and 2, was processed and classified. From this image we digitized the route and morphology of the main stream Medjerda for the year 2013 (Figure 23).

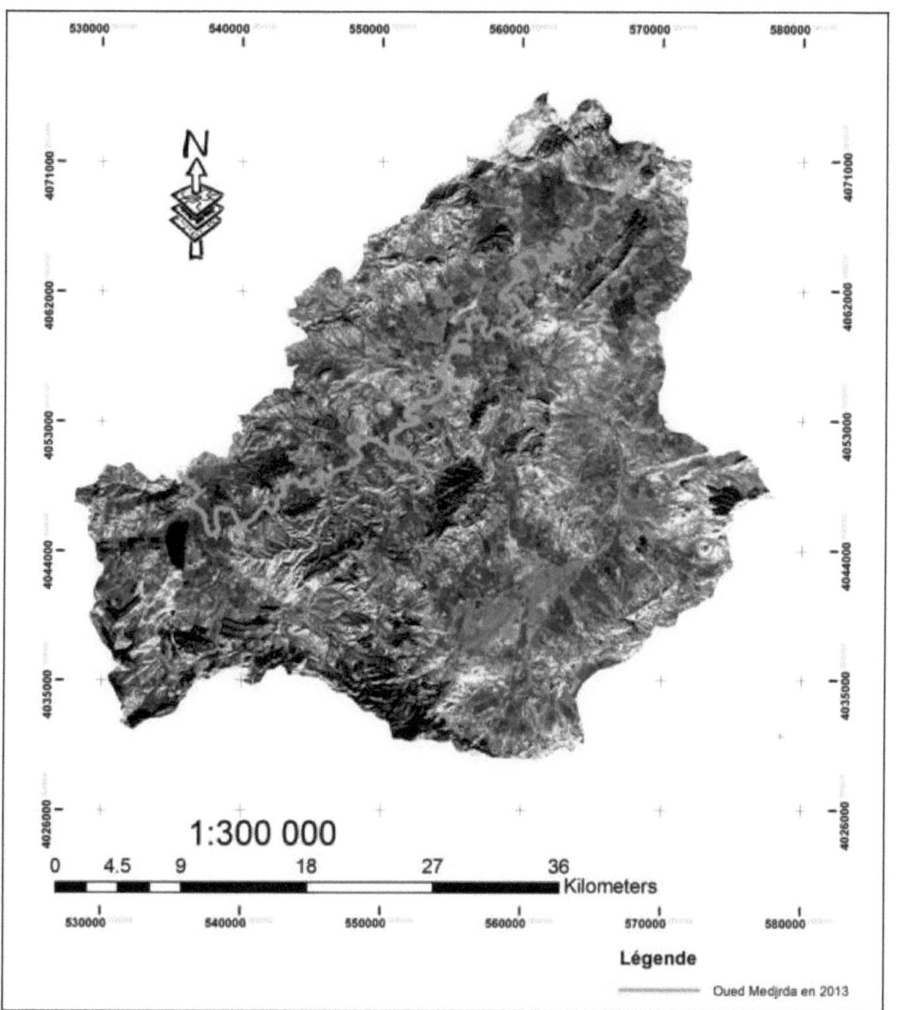

Figure 23. Processed and classified Landsat image (2013) showing the path of the Medjerda river (Sidi Salem-Laaroussia).

4. Evolution of the morphology of oued Mejerda between the years 2003, 2009 and 2013

Following the interpretation of the three Landsat images for the three dates, and the digitizing of the wadi path for these dates, at first sight no differences can be noticed.

However, we proceeded to superimpose the three wadis of different years of acquisition, and we suggested that it would be better for the interpretation of the evolution of Oued Medjerda (section Sidi Salem-Lâaroussia).

Furthermore, by using the satellite image classification tool, we were able to detect the sediment deposition zones on both sides of the bed of the study section. This helped us to justify the different evolutions.

Indeed, the sludge deposited by the wadi are of olive green color more or less pale, homogeneous *(Munsell, 1973), which does* not facilitate the task of classification of sediments. For this, we resorted to the use of aerial photos presented by Google Earth and field visits.

The Sidi Salem-Lâaroussia section is 88 km long, so the general view does not lead us to any valid interpretation. For this, we suggest that the "zoom" tool followed by a scan in the direction of water flow (the red arrow in Figure 24 designates the direction of the scan) are essential during this study (Figure 24).

Since the construction of the Sidi Salem dam in 1981, a rapid rise in the bed of the Medjerda wadi is noted downstream of the Sidi Salem dam. The deposition of solids causes the progressive narrowing of the water passage sections. Indeed, the flow causing the overflow has dropped to less than 200 m^3 /s, a value much lower than the overflow existing before the construction of the dam which was 700 m^3 /s (Rjeb et *al.*, 2003).

- At the Testour level, we notice important alluvial deposits on the right bank of the wadi (Figure 24.a)
- At the Slouguia station, the treatment showed a significant amount of alluviation on the left bank (Figure 24.b)
- At the level of the Medjez el Bab station, all along the meander we note a narrowing of the minor bed of the wadi caused by the important alluviation of the right bank (Figure 24.c)
- Arriving at the Borj Toumi station, we note a weak deposit at the level of two banks (Figure 24.d).

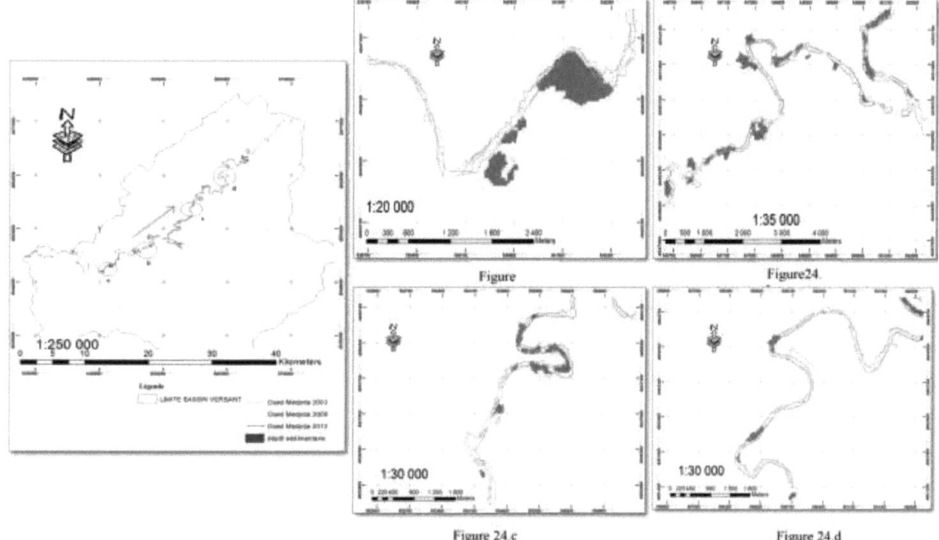

Figure 24. Evolution of the meanders of the middle Medjerda valley as a function of time.

II. Depth index: DI

The depth index (Depth Index) allowed us to evaluate the evolution of the water depth along the Sidi Salem-Lâaroussia section.

It combines visible and red, having the following formula (Brice, 2012):

$$\mathbf{DI = sqrt\ [(VIS)^2 + RED)\ ]^2}$$

Indeed the result provided by the processing of satellite images is not an image but reflectance measurements in specific wavelengths. The calculated indices represent scalars describing the content of the pixels.

The results are shown in Figure 25.

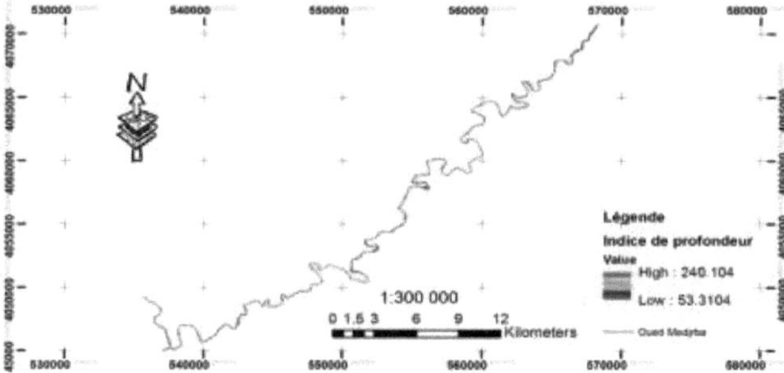

Figure 25 a. Depth index for the year 2003.

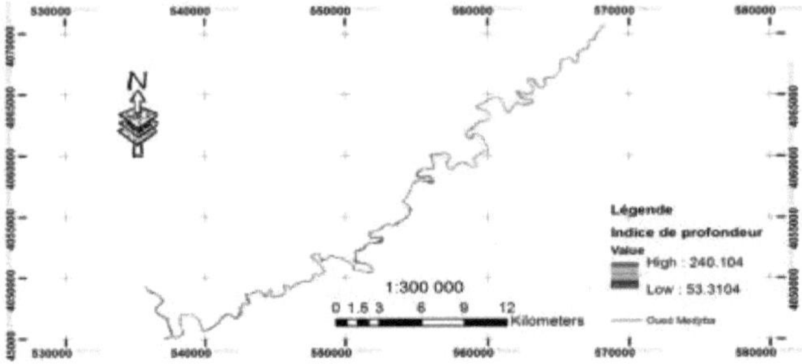

Figure 25 b. Depth index for the year 2009.

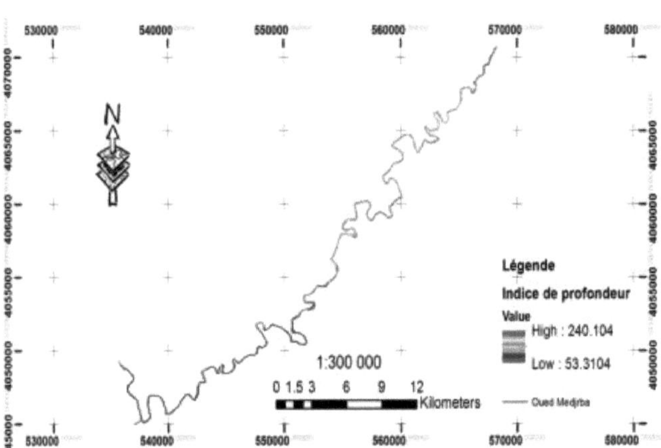

Figure 25 c. Depth index for the year 2013.

Figure 25. Evolution of the depth of the Sidi Salem-Lâaroussia section.

The bottom of the minor bed has undergone between 2003 and 2009 digging at the station Medjez Lbab. As for the level of the Toumi station, we have a continuous digging from the year 2003. Following the various treatments of satellite images of the downstream of the dam Sidi Salem, we noticed :

- From the dam of the Sidi Salem dam to my area of the Matisse meanders, there is a significant amount of alluviation on both banks of the Oued.
- Upstream of the Lâaroussia reservoir, a clear digging is observed.

Indeed, from one release to another, we witness both a phenomenon of alluviation and digging of the wadi.

III. vegetation index: NDVI

At this level, we used the most widely used vegetation index calculation, the NDVI (Normalized Difference Vegetation Index). The regular monitoring of this index will allow us to follow the evolution of the vegetation in a precise area.

It combines the red R and the near infrared PIR, having the following formula (Brice, 2012):

NDVI = (PIR-R)/ (PIR+R)

The extraction of the information is done in two essential steps:

-1- **The calculation of the index**:

It is a process that allows to delineate homogeneous entities or homogeneous areas of the image. It is based on the reflectance of the object. From the resulting image, it is then possible to select the entities corresponding to the different layers without having to redraw the contours.

-2- **classification**:

It is a classification process with supervised learning by object. The learning process consists in manually capturing the vegetation according to the defined themes on tiles representing one or more themes. This learning base can automatically classify the vegetation on request of the operator.

Field verification tours are necessary to confirm certain types that are difficult to identify on screen.

Source: Landsat 7 ETM+ image, dated January 4, 2003. (Figure 26)

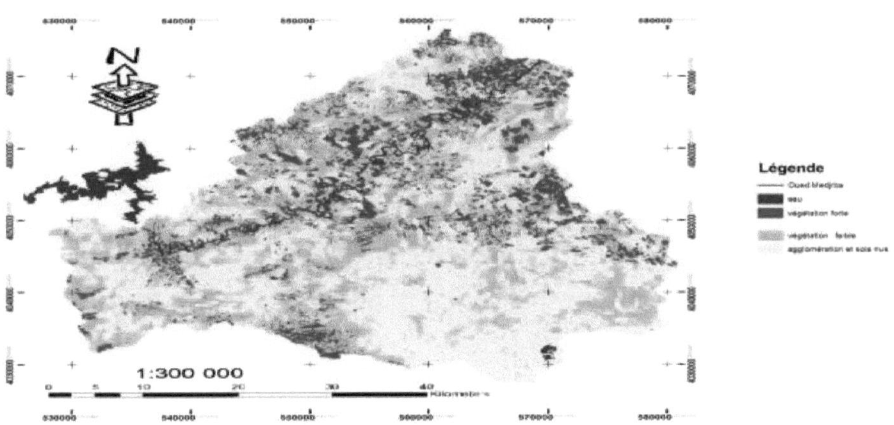

Figure 26. Vegetation index map for the year 2003.

2. Vegetation map of the year 2009 :

Source: Landsat 5 TM, dated November 28, 2009. (Figure 27)

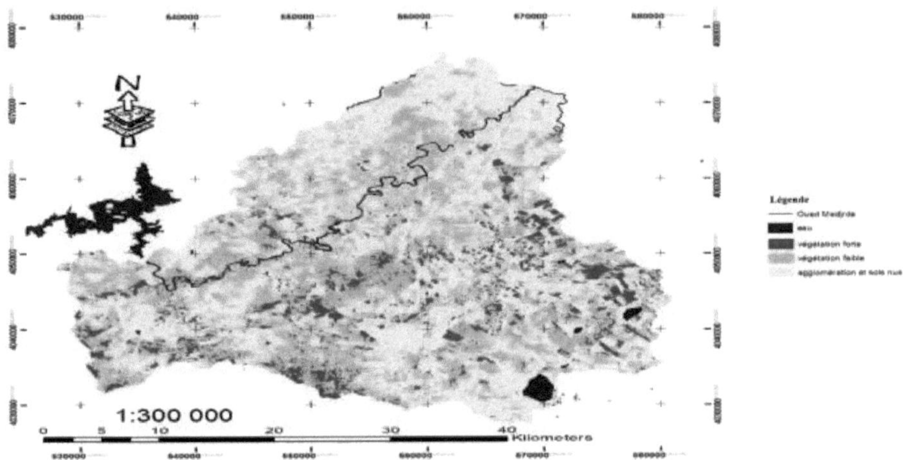

Figure 27. Vegetation index map for the year 2009.

3. 2013 Vegetation Map:

Source: Landsat 8 OLI_TIRS, dated 25 December 2013. (Figure 28)

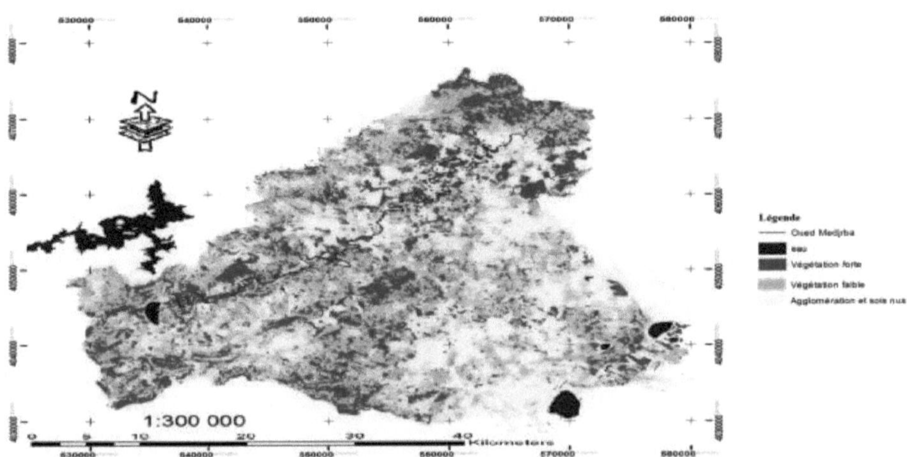

Figure 28. Vegetation index map for the year 2013.

It should be noted that all Landsat images are acquired in the same season, winter.

Indeed, comparing the maps 2003 and 2009, we note a decrease in the density of vegetation on the left bank of the section Sidi Salem - Laaroussia. While the map of the year 2013 presents the densest vegetation.

In the same framework, we note that the vegetation on both sides of the study section shows a remarkable regression. In fact, these areas present not only the most fertile plots but also the most affected by the floods, hence the decrease of the cultivation of these areas.

IV. Armour index: IC

It is considered more efficient than the gloss index to differentiate between built and unbuilt areas. From this index we could determine the roads, the built-up areas and the unbuilt areas.

It combines GREEN and RED, having the following formula (Brice, 2012):

$$IC = 3 * V - R - 100$$

The following maps show the armor index classification results for the years 2003 and 2013. (Figures 29 and 30)

We rely on a diachronic mapping of the built environment in order to better understand the dynamics of space occupation and urbanization. The objective is to establish a map of the extension of built-up areas. To facilitate comparison and evaluation of the progression, disappearance of the built-up area between the dates selected. (Figure 31)

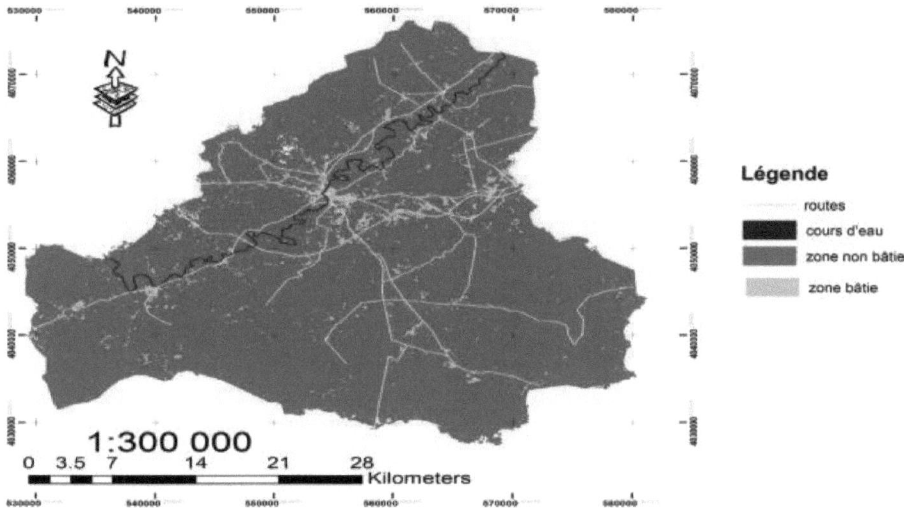

Figure 29. Classification of armor index (built-non-built) for the year 2003.

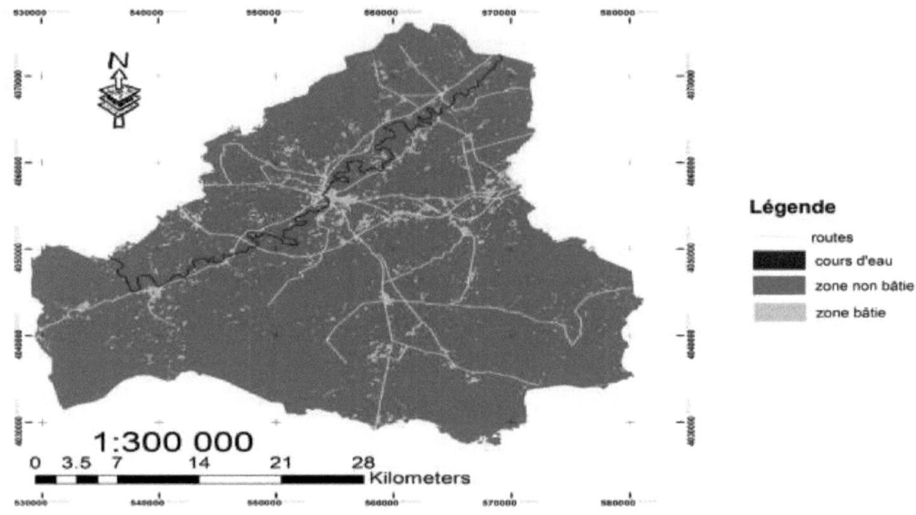

Figure 30. Armor index classification (Built-unbuilt) from the year 2013.

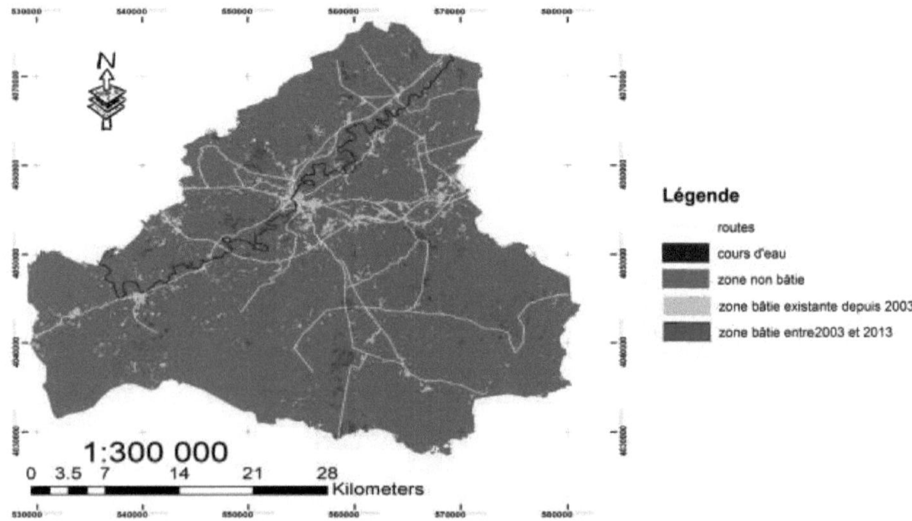

Figure 31. Map of the evolution of the built-up area between 2003 and 2013

This index allowed us to calculate the percentage of the built-up area during the two dates.

$$\% \text{ zone bâtie} = \frac{surface\ de\ la\ zone\ bâtie}{surface\ totale\ du\ bassin\ versant} \times 100$$

En 2003 : $\%$ **zone bâtie** = $\frac{64.154}{1468} = 4.37\ \%$

En 2013 : $\%$ **zone bâtie** = $\frac{82.41}{1468} = 5.61\ \%$

The increase in the built-up area seems to have been not only small (18 km² in 10 years) but also very dispersed. This is not surprising since our study area is a rural area characterized by a low demographic pressure and a great capacity of regeneration of the natural environment.

In the next step, we will map the land cover of the most recent Landsat image we have (2013).

V. Land Use Map:

The land use map is made from the date the most recent Landsat image we have (2013).the study section Sidi Salem- Laaroussia.

- The section of oued Medjerda Sidi Salem- Lâaroussia
- The built-up area (agglomeration and industrial zones)
- The roads
- Andalus Bridge in Medjez Elbab (ALMOURADI)
- Borj Ettoumi Bridge
- Esslouguia Bridge
- Bridge National Road 5
- Field and vegetable crops
- Tree cover
- Olive trees
- Bare floors
- Sedimentary deposits
- Sidi Salem Dam
- Laroussia Dam

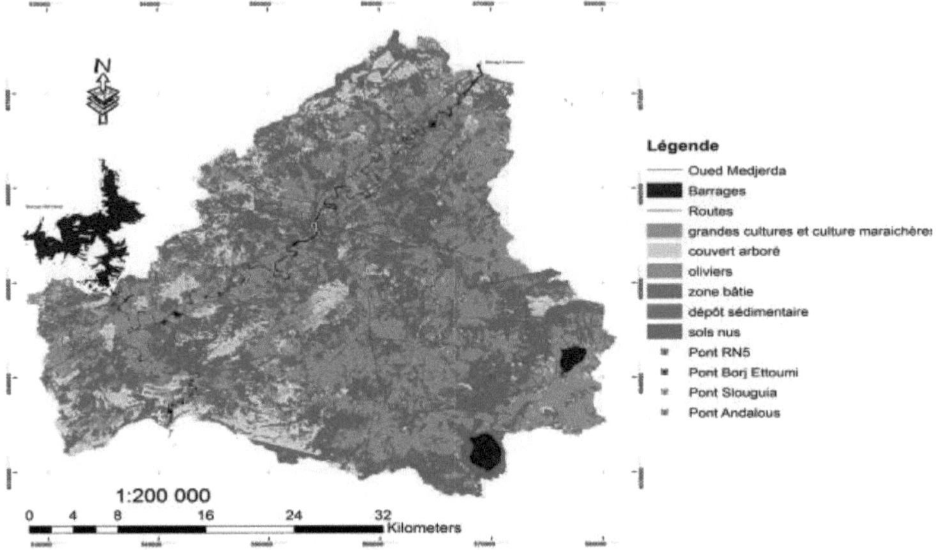

Figure 32. Land use map for the year 2013.

Conclusion

In this chapter, we have demonstrated that from Landsat satellite images we can extract the digital terrain model, delineate the different topographic watersheds of the study area. In addition, we were able to detect the different evolutions of the morphology of the watershed of the middle valley of the Medjerda. The meanders of the section Sidi Salem - Laâroussia were affected in a progressive and increasing way by the sedimentary deposit on both sides of the banks of the minor bed because of the laches of the Sidi Salem Dam. As for the depth of the bed, the digging undergone by the water current is remarkable at the level of the upstream of Laâroussia Dam.

In addition, to be able to establish the land use map, we resorted to the calculation of the armor index, which gave us the evolution of the built-up area over 10 years (between 2003 and 2013) and the vegetation index, which, in turn, facilitated the interpretation of the evolution of the vegetation density over the same period of time.

Chapter III: Hydraulic modeling and flood zone mapping

Introduction

The main objective of this part is the mapping of flood areas of the middle valley of the Medjerda. In this part of the study, a simulation of the floods of the section Sidi Salem-Lâaroussia was carried out using the data of peak flow provided by the regional direction of the water resources, for the identification of the floodable zones. The Arc GIS software was used with the HEC-GeoRAS extension for the pre- and post-processing of the data. The simulation of the wadi flood was established under the hydraulic model HEC-RAS.

I. Basic pretreatment parameters:

1. Section geometry:

This step consists in vectorizing, from the satellite image, first the central section of the water flow, then the minor bed and finally the major bed.

You should also remember a very important rule of digitizing with HEC-GeoRAS: Digitizing is always done from upstream to downstream and from left to right (when looking downstream). Finally, the last rule is that none of your objects must exceed the limits of the TIN (the Digital Terrain Model) extracted, in the first part, from the Landsat image.

2. Creation of cross sections:

The HEC-GeoRAS software stores the cross sections in the 'XSCutLines' layer. One can digitize as much as possible but the closer they are, the more relevant the HEC-GeoRAS analysis will be as well as the final result in Arc GIS. The following integrity constraints must also be scrupulously respected: The cross sections must be perpendicular to the flow direction, they must be wider than the major bed, always digitized from the left to the right (looking downstream) and at the most regular interval possible. Finally, if obstacles (bridges, etc.) to the flow are digitized, a profile must be placed just upstream and another just downstream.

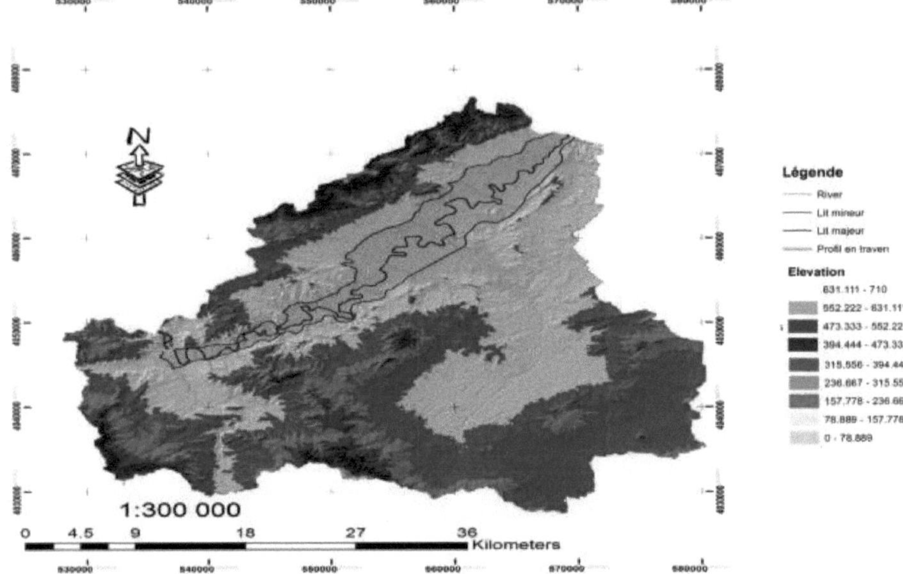

Figure 33. Parameters introduced during preprocessing with HEC Geo-RAS

II. Identification of Hydraulic Works :

The existing bridges on the studied section of the river have to be entered into the software data. The geometrical characteristics of the existing bridges on the study section are collected from the Directorate of Bridges and Roads.

The simulation at these structures requires the following data to be entered:

- Geometric characteristics: Length, Width.
- The bridge model: perched, suspended...
- Hydraulic parameters: expansion coefficient and extraction coefficient...

Table 5. Characteristics of bridges on the study section (Yousfi, 2003).

Bridge	Route	Type	Length	Number of bays
Borj Ettoumi	RL53 1	Beams	160	8
Andalusian	RN5	Vault	122	17
RN5 Bridge	RN5	Beams	112	3
Slouguia	RN5	Beam	140	4

⇒ These bridges are perched bridges.

⇒ The Andalusian bridge presents a case where the transition is abrupt.

1. Roughness coefficients:

Roughness expresses the surface condition of a land. Thus, a heavily vegetated area has a high roughness (low friction coefficient) and the flow is slowed down. However, the bed of a watercourse made up of fine sediments has a low roughness, which favours the flows.

In HEC-RAS, the roughness coefficients are entered as Manning's coefficients, i.e., the inverse of Strickler's coefficients. The values of the initial coefficients were taken from the *HEC-RAS* reference manual. *(Appendix 7)*

These parameters are adjusted by iterative calculations for a historical event, for which the water line is known thanks to the information of the flood marks, so that the model is as representative as possible of the characteristics observed for this event.

Dense urban areas are not directly represented in the model geometry, they will be taken into account by increasing the roughness.

Roughness expresses the surface condition of a land. Thus, a heavily vegetated area has a high roughness (low friction coefficient) and the flow is slowed down. However, the bed of a watercourse made up of fine sediments has a low roughness, which favours the flows.

2. Coefficient of contraction and expansion:

The HEC-RAS software takes the head loss into account when calculating the water line. The widening and narrowing of the channel leads to the expansion and contraction of the flow respectively. This is especially apparent when obstacles and structures are present in the watercourse. The values of the coefficients were taken from the *HEC-RAS* reference manual.

Table 6. Values of contraction and expansion coefficients.

Situation	**Contraction**	**Expansion**
No transition loss to calculate	0	0
Gradual transition	0.1	0.3
Typical bridge section	0.3	0.5
Abrupt transition	0.6	0.8

111. Simulation:

The selected simulation flows are those of the peak periods of the middle Medjerda valley.

These flows vary beyond 220 m^3 /s (the full flow before overflow). So for a recurrent flood of return period of 5 years the flow is 260 m^3 /s, for that of 50 years the flow is 650 m^3 /s and for the exceptional floods of return period of 100 years the flow is equal to 780 m^3 /s (Ousli, Mahjoub 2006)

1. Cushioning:

a. Calibration test:

The calibration phase of the model consists in adjusting the various calculation parameters of the model, especially those of the soil roughness coefficients (Strickler coefficient) and the flow coefficients at the level of the hydraulic structures.

The roughness parameters are initialized according to the land use in the first part of this study. Thus, a roughness coefficient was assigned to each profile by homogeneous sector (major bed, minor bed, urbanized sector, rural sector...).

For this, the calibration test is done by minimizing the relative error.

With E : Relative error

V_{cal} : Calculated value

V_{obs} : Observed value

b. Calibration results:

As indicated above, for the first simulation, we take the following roughness values:

Tableau 7. Values of roughness coefficients for the reference stations.

Station	N left banks	N minor bed	N straight edges
Slouguia	0.075	0.02	0.075
Medjez Lbab	0.075	0.0675	0.075
El Herri	0.075	0.0675	0.75
Borj Ettoumi	0.075	0.06	0.075

By varying the value of the roughness of the different sections, the minimum error corresponds to the values of the roughness given by the following table:

Tableau 8. Values of calibration roughness coefficients for the reference stations.

Station	N left banks	N minor bed	N straight edges

Slouguia	0.075	0.027	0.075
MEdjez Lbab	0.075	0.0675	0.075
El Herri	0.09	0.07	0.09
Borj Ettoumi	0.085	0.055	0.085

The results of the water level calibration at different stations for the flows 260 m^3 /s, 650 m^3 /s and 780 m^3 /s are shown in the following table:

Table 9. Setting the water level

Slouguia			
	Calculated height (m)	Observed height (m)	Error
260 m /s^3	6.2	6.68	-7.18562874
650 m /s^3	6.6	6.8	-2.94117647
780 m /s^3	7.5	7.7	-2.5974026
Medjez El Bab			
	Calculated height (m)	Observed height (m)	Error
260 m /s^3	7.24	7.59	-4.6113307
650 m /s^3	7.4	7.8	-5.12820513
780 m /s^3	8.2	8.5	-3.52941176
El Herri			
	Calculated height (m)	Observed height (m)	Error
260 m /s^3	5.51	5.5	0.18181818
650 m /s^3	5.9	5.8	1.72413793
780 m /s^3	6.8	6.8	0
Borj Ettoumi			
	Calculated height (m)	Observed height (m)	Error
260 m /s^3			
650 m /s^3	4.2	4.6	-8.69565217
780 m /s^3			

⇒ The error in absolute value is between 0% and 9%.

We note that the heights calculated at Slouguia, Medjez Elbab, El Herri and Borj Ettoumi, are underestimated. The difference is, in fact, acceptable.

⇒ The distribution of roughness presents values varying from 0.027 to 0.09. This variation is explained by the variation of the wadi bed cover and the diversity of vegetation and its density.

2. Results:

The HEC-RAS software allows to present the results of the simulations in different forms such as the profile of the water line calculated for the chosen simulation flow (figure 34), with the different hydraulic parameters.

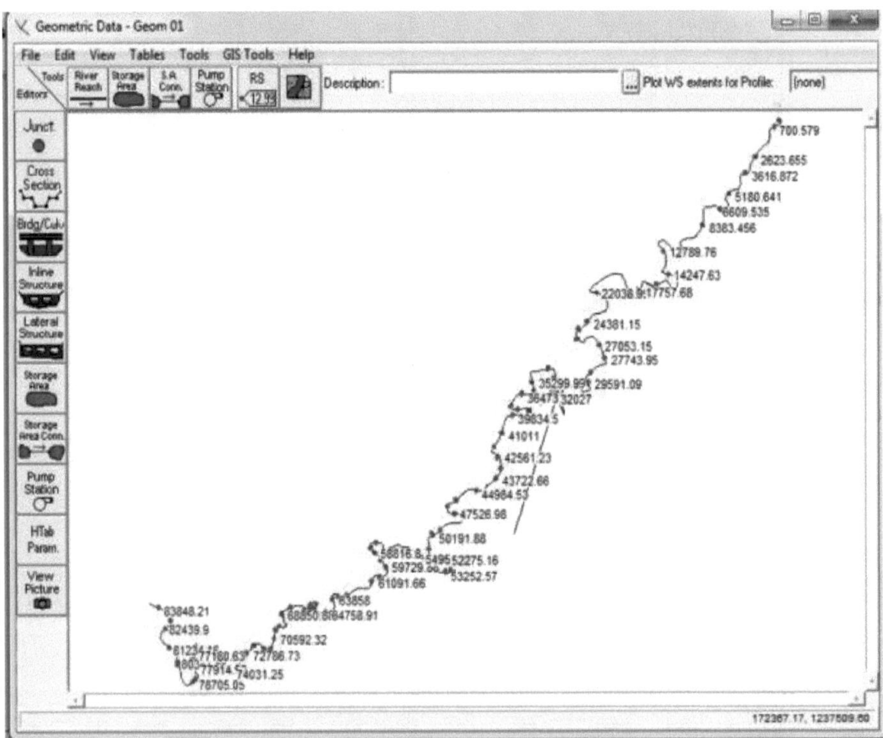

Figure 34. Profile presentation of the study reach on HEC-RAS.

a. Hydraulic parameters:

The results are visualized in tables (Appendix 8), where we find mainly the minor bed elevation, the water surface elevation, the hydraulic slope, the velocity and the Froude number for each section.

The table in the appendix shows that the speeds decrease between Slouguia and Medjez Elbab. This decrease favors the deposition of suspended solids especially at the level of the bridges.

b. The profile of the water line

The results are presented in the following table:

Table 10. Variation in water level.

Q (m^3 /s)	H $_{Slouguia}$ (m)	H $_{Medjez\ Elbab}$ (m)	H $_{Herri}$ (m)
260	6.2	7.24	5.51
650	6.6	7.4	5.9
780	7.5	8.2	6.8

Going from upstream to downstream of the section Slouguia Medjez Elbab, we note that for the same flow, the height increases slightly. While in EL Herri, the water height decreases.

Indeed, it is estimated that this is due to the existence of three bridges Slouguia, Andalous (Mouradi) and that of the RN5 which are responsible for a large loss of load between Medjez Elbab and Slouguia.

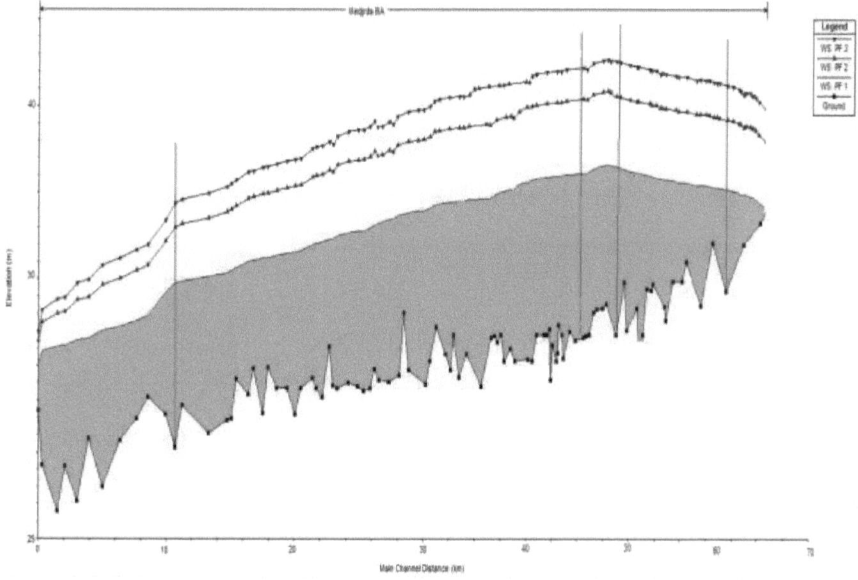

Figure 35. Profile of the water line of the Sidi Salem- Laaroussia section

IV. Postprocessing on HEC Geo-RAS:

Thanks to the Geo-Ras tool (Ras Mapping) we can extract flood maps by surface and water height (Innundation Mapping) and import them into Arc GIS.

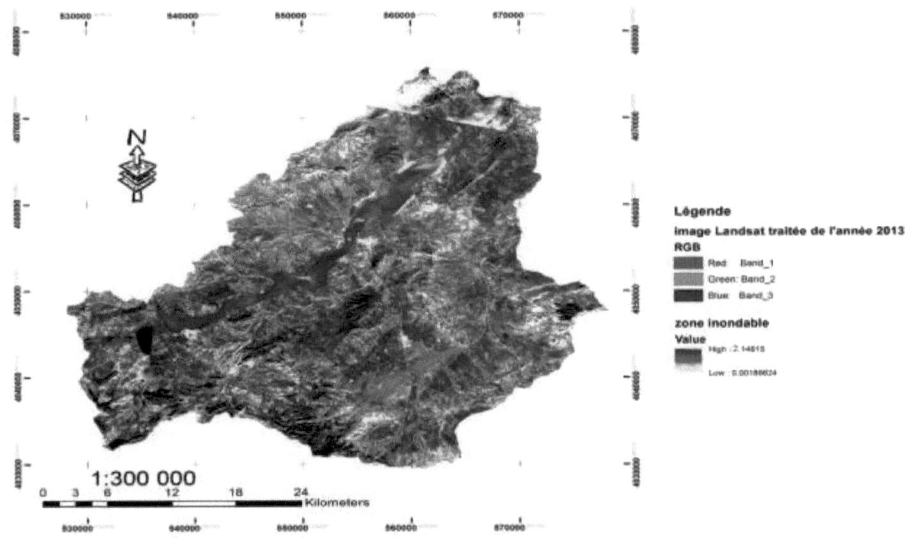

Figure 36. Map of flood zones and estimated water height.

From the following map (Figure 36) we could conclude that :

J The flooding begins at the meandering El Matisse area, about 10 km downstream Medjez-El-Bab city.

J The water depth reached more than two meters in some places.

J At Medjez-El-Bab, the overflows start from a flow of 700 m3 / s.

J The flow regime is fluvial throughout and partly explains the intensity of sediment deposition.

J sediment erosion and deposition mainly explain the new distribution of the Manning's coefficient and the real need for a new approach taking into account the morpho-dynamics of the river bed.

Conclusion

In this last but one chapter, we were able to determine the flood zones and the different water heights from the hydraulic simulation for three return periods 5 years, 50 years and 100 years.

This is an important part of the mapping of flood risk areas in the next chapter.

Chapter IV: The Flood Risk Map

Introduction

The risk is therefore considered as a measure of the dangerous situation that results from the confrontation of the hazard and the stakes. This measure is often expressed in terms of severity and probability and, as for technological risk, can be represented in the Farmer diagram.

Indeed, and as indicated in the first part (chapter I,$II) The usual formula given for the natural risk is the following:

(Risk) = (hazard) x (issue)

I. Issues:

Today, most rivers have a completely different geometry than that shaped by past floods. The minor bed has often been rectified, dammed or re-calibrated. The major bed has been urbanized and transportation infrastructures have been built in the flood expansion fields. The physical limits of flow that constituted the high terraces have been replaced by dykes and road or railroad embankments. However, an exceptional flood may occur which is capable of submerging the structures present in the major bed. Therefore, it is important to take into account the areas that seem to be protected, but that could be affected by an exceptional flood. (Kreis, 2004)

All issues located in this area that may be subject to damage from a hazard are designated as elements at risk.

In this study, the flood risk assessment will focus on urban areas, infrastructures and cultivated areas. For this purpose, maps representing built-up areas, road networks and the distribution of cultivated areas will be considered.

These issues were deduced from the existing land use map (Figure 32), while adding existing developments in the study area.

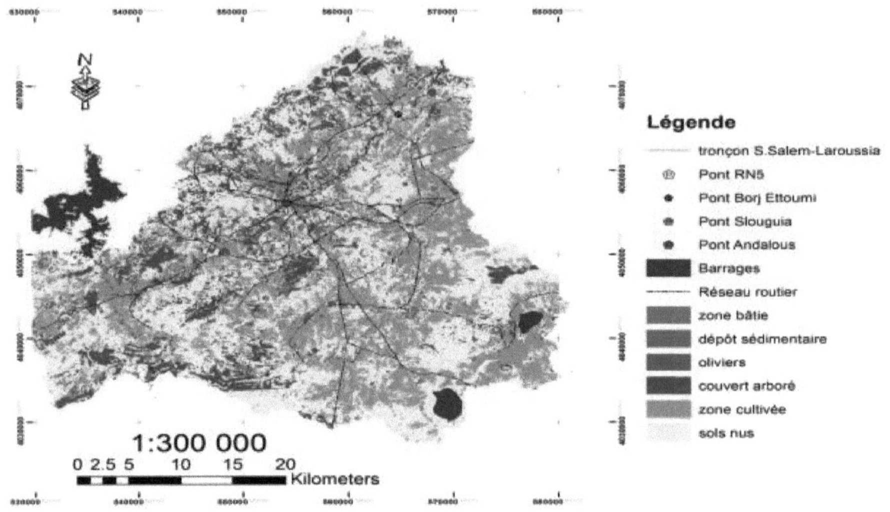

Figure 37. Map of issues exposed to flooding.

II. Flood hazard:

The hazard is a physical phenomenon, natural and uncontrollable, of given occurrence and intensity. It can be characterized according to two components, one frequent (occurrence), the other spatio-temporal (intensity).

Indeed, the term hazard refers to areas susceptible to more or less significant and frequent flooding, following the natural overflow of a watercourse. The following map delineates areas that are characterized by one of the two proposed hazard values: low and high.

These values were deduced from the flood extent map and the water level illustrated in the previous chapter (Figure 36).

A strong hazard means, in our case, an extensive flooding that occurs during recurrent events and with a water height more or less high. As for the weak hazard, it means the limit of flooding during extreme and rare events.

The simple superposition of the hazards with the land use map showed us, in a simple and visual interpretation, the potentially flooded areas.

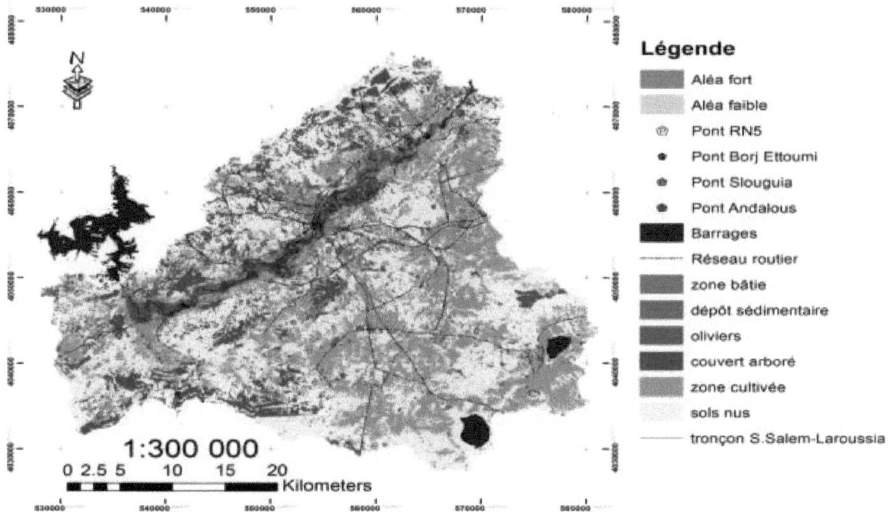

Figure 38. Hazard map.

III. The flood risk map.

Risk is the result of a complex hazard combined with vulnerability (A. DAUPHINÉ, 2001).

For the purposes of this study, the flood risk map is the product of the two maps shown above, the issues map and the hazard map.

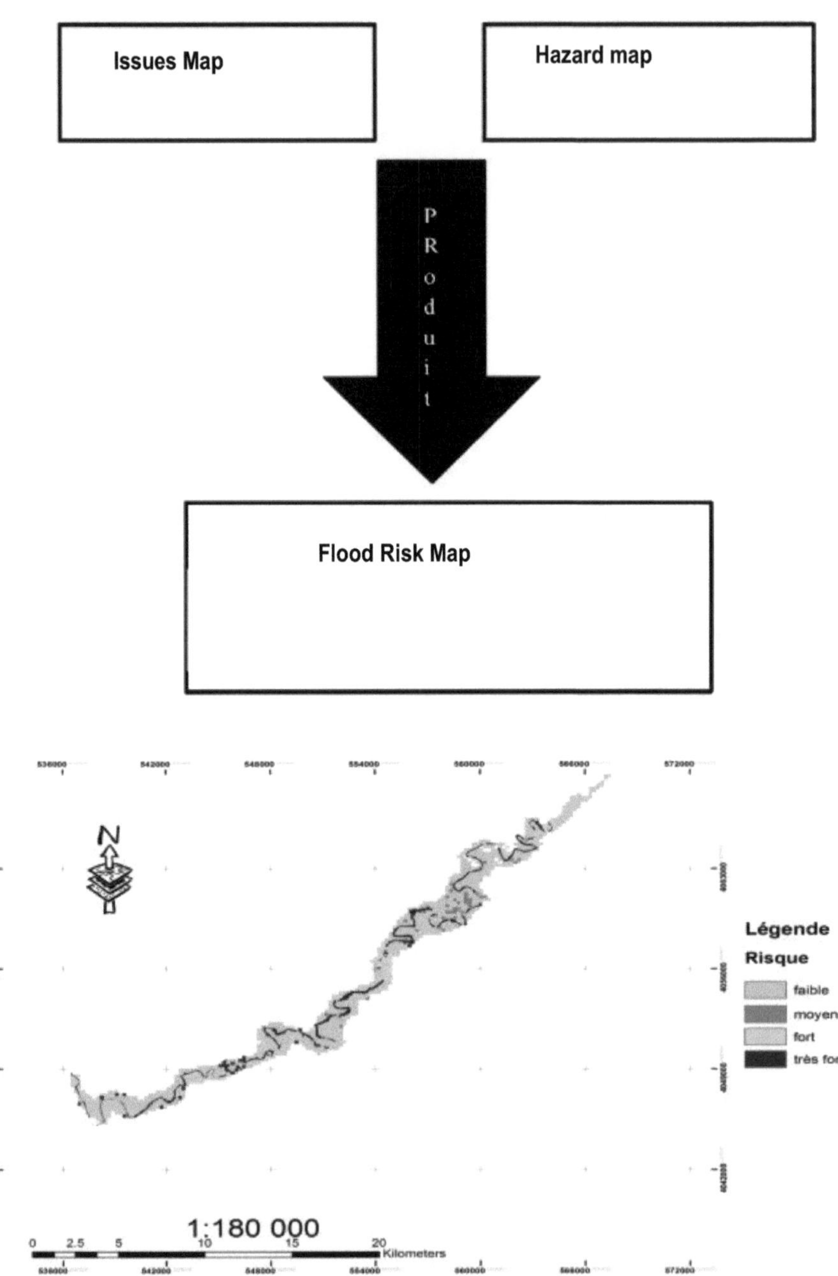

Figure 39. Map of flood hazard areas.

The final map obtained allowed us to detect the areas at risk of flooding according to four selected values of risk levels (low, medium, high and very high). For a frequent event (annual or bi-annual rainfall) the areas with high flood risk (in blue) are the first to be affected. With the same approach, the low risk areas (in pink) will only be affected during an extreme event (100-year rainfall).

The study area appears to be at high risk of flooding in the alluvial plains adjacent to the minor and major beds.

This map showed us that at the level of the bridge of Slouguia the risk of flooding is very strong (dark blue color). The justification of this high level of risk is not limited to the presence of this bridge which presents an obstacle for the flow of water during a flood but also, it is justified by the increasing quantity of alluvial deposits on both sides of the minor bed of the section of the Wadi (chapter I of the second part).

Similarly, the flood risk map above confirmed that Medjez Elbab has a high risk of flooding (orange color). This level of risk is justified by the fact that the area has a high agglomeration compared to other areas, hence the decrease in water infiltration and the increase in the time of waterlogging. It should be noted that this study does not take into account the overflow of the sewerage systems of the urban areas. Concerning the Matisse meander, it presents a medium level of flood risk. This risk is, in fact, justified by considering this zone as an area with a high water saturation rate. It is a meander with a sedimentary deposit more or less high and which does not stop increasing. It is true that it is a fertile area and a favorable field for crops but it is also exposed to frequent flooding.

Conclusion

In this chapter, we were able to determine the different classes of flood risk, and we were able to detail the areas exposed to them.

Faced with a given natural hazard, society must answer two fundamental questions:

- What level of protection is desired?
- What level of risk can be accepted?

General conclusion

The objective of this work is the interpretation of satellite images to determine the evolution of the morphology of the catchment area of the Middle Medjerda Valley following the major floods of 2003, 2008 and 2011 and to establish the map of the extent of flooding and the height of water followed by the development of the flood risk map. This reach must then be modeled in order to be able to simulate and test the possible scenarios by means of one-dimensional hydraulic modeling.

The driving idea is then to extract the maximum data from the available satellite images, and evaluate any change detected. This study allowed to determine the morphological evolution of the study section of the middle Medjerda valley such as the evolution of the wadi bed depth, meanders, solid contributions.

In addition, the interpretation of satellite images allowed us to study the land use and its evolution over time.

By integrating the remote sensing results into the one-dimensional HEC-RAS model, we were able to simulate floods starting at a flow rate of 220 m^3 /s (full-tide flow before overflow).

The calibration parameter of the HEC-RAS model is the roughness coefficient (n). The values of Manning (n) vary between 0.02 and 0.04 in the minor bed and between 0.062 and 0.085 in the banks.

Scenarios of constant flow releases (300 m^3 /s, 650 m^3 /s, 780 m^3 /s) from Sidi Salem dam have been tested. The results obtained show that we begin to observe the relative overflow from 300 m^3 /s, accentuated with a flow of 650 m^3 /s. The overflowing sections are concentrated mainly in the city of Medjez El Bab, at the level of the Elmouradi bridge (AlAndalous) and the meander Matisse, El Herri and Borj Ettoumi which require protection against similar events that may occur.

High resolution satellite images such as quickbird and Spot could be more reliable to have a more adequate degree of certainty. But this study can be a reference for studies to reduce the risk of flooding in the most affected areas.

Bibliographic list

Ali A. & Qadir D. A., 1989, Study of river flood hydrology in Bangladesh with AVHRR data, *International Journal of Remote Sensing,* Vol. 47, pp. 1873-1891.

Amara A, 2012, Etude de la dynamique fluviale de la Medjerda dans le tronçon Sidi Salem Laâroussia, PFE, ESIER.

Ancey C., 2005, Une introduction à la Dynamique des Avalanches et des Écoulements Torrentiels, *Course*, Ecole Polytechnique Fédérale de Lausanne, http://www.toraval.fr/articlePDF/intro-risk.pdf .

Azizi N, 2008, characterization of sediments and study of the mechanisms of solid transport in suspension of the Medjerda, case study: Slouguia- Elherri section, Master, ESIER.

Bates P. D., Horritt M. S., Smith C. N. & Mason D., 1997, Integrating remote sensing observations of floodhydrology and hydraulic modelling, *Hydrological Processes,* Vol. 11 (14), pp. 1777-1795.

Bates P. D., Marks K. J. & Horritt M. S., 2003, Optimal use of high-resolution topographic data in flood inundation models, *Hydrological Processes,* Vol. 17 (3), pp. 537-557.

Bonn F. & Rochon G., 1993, Précis de Télédétection - Vol.1 Principes et méthodes, 485.

Bukata R. P., Jerome J. H., Kondratyev K. Y. & Pozdniakov D. V., 1995, Optical Properties and Remote Sensing of Inland and Coastal Waters, *Book*, 362 pp.

Brice ANSELM (2012), Digital processing of satellite images; multispectral transformation.

CCRS@Canada Centre for Remote Sensing, *website*, http://ccrs.nrcan.gc.ca/.

Campbell, J.B. (1987) Introduction to Remote Sensing. The Guilford Press, NewYork).

Casenave A. & Valentin C. (1989) - Les états de surface de la zone sahélienne. Ed. ORSTOM, Collection Didactiques, 227 p.

Commission interministérielle de terminologie de la télédétection aérospatiale, 1988.

Clanzig Stéphane and Yann Chancelier, 2010, La géologie de la Tunisie, Lycée Flaubert, Litho thèque Tunisienne. www.lyceeflaubert-lamarsa.com/lithotheque.

Claude J, Rodier JA, Colombani J and Kallel R, 1981, Le bassin de la Medjerdah, 472pp.

DAUPHINÉ A. (2001) - Risques et catastrophes : observer, spatialiser, comprendre, gérer.

Édit. Armand COLIN, Paris, 287 p.

Dhakal A. S., Amada T., Anyia M. & Sharma R. R., 2002, Detection of areas associated with flood and erosion caused by a heavy rainfall using multitemporal Landsat TM data, *Photogrammetric engineering and remote sensing,* Vol. 68, pp. 233-240.

Giacomelli A., Mancini M. & Rosso R., 1995, Assessement of flooded areas from ERS-1 PRI data: An application to the 1994 flood in Northern Italy, *Physics and Chemistry of The Earth,* Vol. 20 (5-6), pp. 469-474

Hechmet H, 1989, CARTE DES RESSOURCES EN EAU DE LA TUNISIE AU 1/200.000 Feuille de Tunis n°5 ; Notice explicative.

Horritt M. S. & Bates P. D., 2002, Evaluation of 1D and 2D numerical models for predicting river flood inundation, *Journal of Hydrology,* Vol. 268 (1-4), pp. 87-99.

Hudson P. F. & Colditz R. R., 2003, Flood delineation in a large and complex alluvial valley, lower Panuco basin, Mexico, *Journal of Hydrology,* Vol. 280 (1-4), pp. 229-245.

Imhoff M. L., Vermillion C., Story M. H., Choudhury A. M., Gafoor A. & Polcyn F., 1987, Monsoon flood boundary delineation and damage assessment using space borne imaging radar and Landsat data, *Photogrammetric Engineering and Remote Sensing,* Vol. 47, pp. 405-413.

Joveniaux S., 1986, Etude et suivi des inondes : utilisation de la télédétection spatiale, *Report*, Institut de Mécanique des fluides de Strasbourg.

Kreis N, 2004, Modelling floods of mid-mountain rivers for integrated flood risk management, 350 pp.

Lillesand, T.M. and Kiefer, R.W. (1994) Remote Sensing and Image Interpretation. John Wiley and Sons Inc, New York.

Lamachère J.M. et Puech C. (1995) - Remote sensing and regionalization of soil runoff and infiltration in Sahelian and North Sudanian Africa. In: Regionalisation en hydrologie, application au développement; ed. scient. L. Le Barbé and E. Servat. Proceedings of the Villes journées hydrologiques de l'ORSTOM, Montpellier, 22-23 September 1992; ORSTOM Editions, colloques et séminaires: 205-228.

Maurel P., 2001, Bases de la télédétection, Course, DEA SEEC.

MEDD-PRIM@Portail de la prévention des risques majeurs, Ministry of Ecology and Sustainable Development, website, http://www.prim.net/.

Moussa M. & Laranier R., 2004, Apport des systèmes d'information géographique et de la télédétection à l'analyse du risque d'inondation dans la ville de Saint-Louis du Sénégal, Actes de colloque, Géo risques et télédétection, Ottawa, Canada, pp. 139-141.

Ochir Altansukh,2012, Tuul River and Its Catchment Area Delineation from Satellite Image.

Puech C. & Vidal A., 1995, Proceedings, CEMAGREF-FAO expert consultation on the Use of Remote Sensing Techniques in Irrigation and Drainage, FAO water reports, 4, Montpellier, pp. 151-154.

Rango A. & Anderson A. T., 1974, Flood hazard studies in the Mississippi River basin using remote sensing, Water Resources Bulletin, Vol. 47, pp. 1060-1081.

Russ, John C. (1995) The Image Processing Handbook. 2nd edition. CRC Press, Baca Raton.

Smith L. C., 1997, Satellite remote sensing of river inundation area, stage and discharge: a review, Hydrological processes, Vol. 11, pp. 1427-1439.

Smith L. C., 1997, Satellite remote sensing of river inundation area, stage and discharge: a review, Hydrological processes, Vol. 11, pp. 1427-1439.

Torterotot J. P., 1993, Le coût des dommages dus aux inondations : Estimation et analyse des incertitudes, Thèse de doctorat, spécialité Sciences et Techniques de l'Environnement, Ecole Nationale des Ponts et Chaussées, 284 p. + annexes.

Tholey N., Clandillon S. & DeFraipont P., 1997, the contribution of spaceborne SAR and optical data in monitoring flood events: Examples in northern and southern France, Hydrological Processes, Vol. 11 (10), pp. 1409-1413.

Tralli D. M., Blom R. G., Zlotnicki V., Donnellan A. & Evans D. L., 2005, Satellite remote sensing of earthquake, volcano, flood, landslide and coastal inundation hazards, Isprs Journal of Photogrammetry and Remote Sensing, Vol. 59 (4), pp. 185-198.

USGS@Manning n reference online: Verified Roughness Characteristics of Natural Channels, website,

http://wwwrcamnl.wr.usgs.gov/sws/fieldmethods/Indirects/nvalues/index.htm.

USACE, 2005a, HEC RAS Hydraulic Reference Manual.

USACE, 2005b, HEC RAS User's Manual.

Vidal J.-P., 2005, Validation opérationnelle en hydraulique fluviale - Approche par un système à base de connaissance, Thèse de doctorat, Institut National Polytechnique de Toulouse, HHLY-Cemagref, 303 p. + annexes.

Yousfi A, 2003, simulation of the Medjerda river flows downstream of the Sidi Salem dam and flood mapping, PFE, INAT.

Annexes

Appendix 1. Example of radiometric corrections.

⇒ Landsat ETM+ image of the year 2013

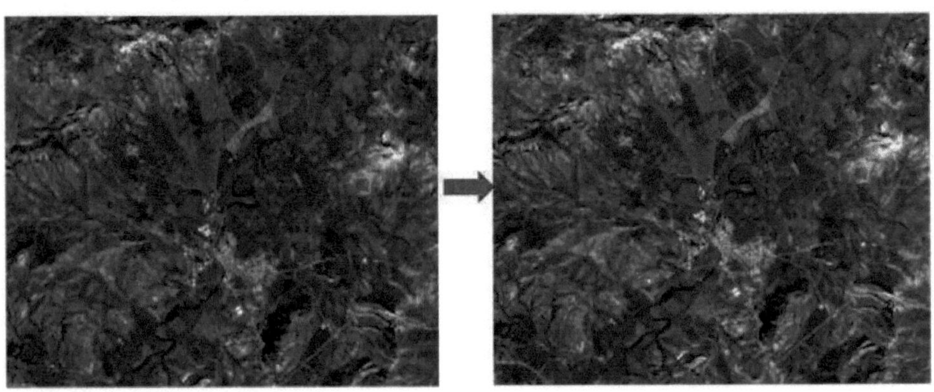

Appendix 2. Example of Streshing.

⇒ Landsat ETM+ image of the year 2013

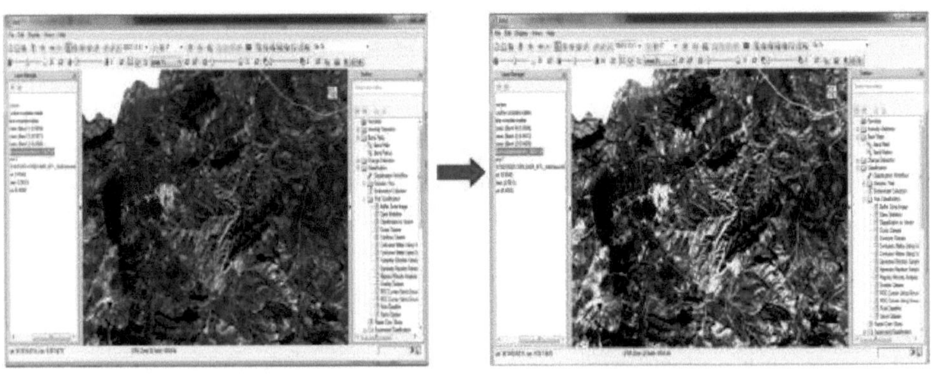

Appendix 3. Hypsometric curve

Altitude (CN) (m)	% of area Cumulative
45	100
50	93.81
100	76.66
150	55.82
200	33.54
250	19.6
300	11.83
350	7.56
400	4.8
450	2.5
500	1.08
550	0.46
600	0.18
650	0.03
700	0

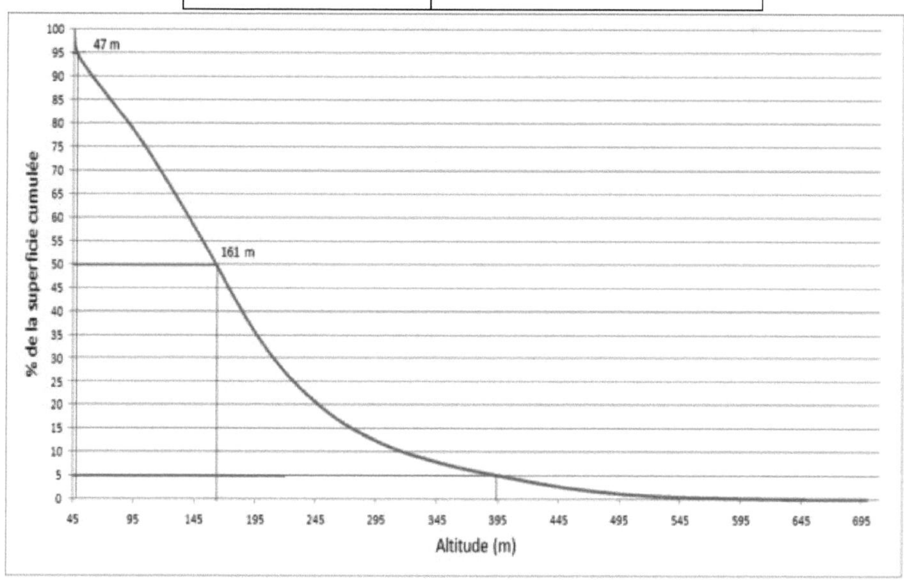

Appendix 4. Classification table of correspondence between basin shape and Kp values.

Interval of Kp	*Watershed shape*
1.00 à 1.25	circular to fairly elongated
1.25 à 1.50	fairly long to long
1.50 à 1.75	elongated to very elongated

Appendix 5. Table of relief classification according to the Ig index.

very low relief	Ig <0.002
low relief	0.002 <Ig<0.005
low relief	0.005 <Ig<0.01
moderate relief	0.01 <Ig <0.02
fairly strong relief	0.02 <Ig <0.05
strong relief	0.05 <Ig <0.1
very strong relief	Ig>0.1

Appendix 6. Emberger's classification table.

Bioclimatic floors	**Q**
Hyperhumid	Q > 170
Wet	110 < Q < 170
Subhumid	70 < Q < 110
Upper semi-arid	50 < Q < 70
Lower semi-arid	30 < Q < 50
Arid	10 < Q < 30
Saharan	Q < 10

Appendix 7. Manning's coefficient from the HEC-RAS manual.

Type of Channel and Description	Minimum	Normal	Maximum
A. *Natural Streams*			
1. **Main Channels**			
a. Clean, straight, full, no rifts or deep pools	0.025	0.030	0.033
b. Same as above, but more stones and weeds	0.030	0.035	0.040
c. Clean, winding, some pools and shoals	0.033	0.040	0.045
d. Same as above, but some weeds and stones	0.035	0.045	0.050
e. Same as above, lower stages, more ineffective slopes and sections	0.040	0.048	0.055
f. Same as "d" but more stones	0.045	0.050	0.060
g. Sluggish reaches, weedy. deep pools	0.050	0.070	0.080
h. Very weedy reaches, deep pools, or floodways with heavy stands of timber and brush	0.070	0.100	0.150
2. **Flood Plains**			
a. Pasture no brush			
1. Short grass	0.025	0.030	0.035
2. High grass	0.030	0.035	0.050
b. Cultivated areas			
1. No crop	0.020	0.030	0.040
2. Mature row crops	0.025	0.035	0.045
3. Mature field crops	0.030	0.040	0.050
c. Brush			
1. Scattered brush, heavy weeds	0.035	0.050	0.070
2. Light brush and trees, in winter	0.035	0.050	0.060
3. Light brush and trees, in summer	0.040	0.060	0.080
4. Medium to dense brush, in winter	0.045	0.070	0.110
5. Medium to dense brush, in summer	0.070	0.100	0.160
d. Trees			
1. Cleared land with tree stumps, no sprouts	0.030	0.040	0.050
2. Same as above, but heavy sprouts	0.050	0.060	0.080
3. Heavy stand of timber, few down trees, little undergrowth, flow below branches	0.080	0.100	0.120
4. Same as above, but with flow into branches	0.100	0.120	0.160
5. Dense willows, summer, straight	0.110	0.150	0.200

Type of Channel and Description	Minimum	Normal	Maximum
3. **Mountain Streams, no vegetation in channel, banks usually steep, with trees and brush on banks submerged**			
a. Bottom: gravels, cobbles, and few boulders	0.030	0.040	0.050
b. Bottom: cobbles with large boulders	0.040	0.050	0.070

Type of Channel and Description	Minimum	Normal	Maximum
B. *Lined or Built-Up Channels*			
1. **Concrete**			
a. Trowel finish	0.011	0.013	0.015
b. Float Finish	0.013	0.015	0.016
c. Finished, with gravel bottom	0.015	0.017	0.020
d. Unfinished	0.014	0.017	0.020
e. Gunite, good section	0.016	0.019	0.023
f. Gunite, wavy section	0.018	0.022	0.025
g. On good excavated rock	0.017	0.020	
h. On irregular excavated rock	0.022	0.027	
2. **Concrete bottom float finished with sides of:**			
a. Dressed stone in mortar	0.015	0.017	0.020
b. Random stone in mortar	0.017	0.020	0.024
c. Cement rubble masonry, plastered	0.016	0.020	0.024
d. Cement rubble masonry	0.020	0.025	0.030
e. Dry rubble on riprap	0.020	0.030	0.035
3. **Gravel bottom with sides of:**			
a. Formed concrete	0.017	0.020	0.025
b. Random stone in mortar	0.020	0.023	0.026
c. Dry rubble or riprap	0.023	0.033	0.036
4. **Brick**			
a. Glazed	0.011	0.013	0.015
b. In cement mortar	0.012	0.015	0.018
5. **Metal**			
a. Smooth steel surfaces	0.011	0.012	0.014
b. Corrugated metal	0.021	0.025	0.030
6. **Asphalt**			
a. Smooth	0.013	0.013	
b. Rough	0.016	0.016	
7. **Vegetal lining**	0.030		0.500

Type of Channel and Description	Minimum	Normal	Maximum
C. *Excavated or Dredged Channels*			
1. Earth, straight and uniform			
a. Clean, recently completed	0.016	0.018	0.020
b. Clean, after weathering	0.018	0.022	0.025
c. Gravel, uniform section, clean	0.022	0.025	0.030
d. With short grass, few weeds	0.022	0.027	0.033
2. Earth, winding and sluggish			
a. No vegetation	0.023	0.025	0.030
b. Grass, some weeds	0.025	0.030	0.033
c. Dense weeds or aquatic plants in deep channels	0.030	0.035	0.040
d. Earth bottom and rubble side	0.028	0.030	0.035
e. Stony bottom and weedy banks	0.025	0.035	0.040
f. Cobble bottom and clean sides	0.030	0.040	0.050
3. Dragline-excavated or dredged			
a. No vegetation	0.025	0.028	0.033
b. Light brush on banks	0.035	0.050	0.060
4. Rock cuts			
a. Smooth and uniform	0.025	0.035	0.040
b. Jagged and irregular	0.035	0.040	0.050
5. Channels not maintained, weeds and brush			
a. Clean bottom, brush on sides	0.040	0.050	0.080
b. Same as above, highest stage of flow	0.045	0.070	0.110
c. Dense weeds, high as flow depth	0.050	0.080	0.120
d. Dense brush, high stage	0.080	0.100	0.140

Appendix 8. Parameters calculated in HEC-RAS during the simulations.

Appendix 8.1. Parameters calculated in HEC-RAS during the simulation for a flow rate of 260 m /s.^{3}

Reach	River Sta	Q Total	Min Ch El	W.S. Elev	E.G. Elev	E.G. Slope	Vel Chnl	Flow Area	Top Width	Froude # Chl
		(m3/s)	(m)	(m)	(m)	(m/m)	(m/s)	(m2)	(m)	
BA	83848.21	260	19.81	27.6	27.65	0.000433	1.15	320.49	50.27	0.14
BA	82974.85	260	21.01	27.43	27.5	0.000722	1.28	275.34	55.64	0.17
BA	82439.9	260	19.86	27.37	27.41	0.000354	1.06	350.32	47.88	0.12
BA	81234.16	260	19.86	27.18	27.24	0.000502	1.25	298.73	42.55	0.15
BA	80343.55	260	17.57	27.1	27.13	0.000263	1.03	378.75	47.51	0.11
BA	80182.37	260	18.42	27.09	27.12	0.000243	0.92	408.57	55.25	0.1
BA	78705.05	260	19.71	26.92	26.97	0.000446	1.14	324.08	47.47	0.14
BA	78479.98	260	19.36	26.88	26.94	0.000495	1.25	297.31	41.51	0.15
BA	77914.52	260	19.43	26.82	26.86	0.000345	0.98	358.46	50.82	0.12
BA	77400.86	260	16.76	26.71	26.78	0.000583	1.5	264.29	32.23	0.16
BA	77180.63	260	12.73	26.73	26.75	0.000117	0.83	463.16	38.45	0.07
BA	76655.31	260	18.31	26.68	26.72	0.000353	1.15	342.56	43.72	0.13
BA	75371.32	260	17.02	26.58	26.61	0.000209	0.94	424.57	53.53	0.1
BA	75059.63	260	19.81	26.52	26.58	0.000696	1.09	276.11	40.63	0.13
BA	74031.25	260	16.76	26.49	26.5	0.000088	0.62	610.48	74.71	0.06
BA	72786.73	260	18.57	26.37	26.43	0.000553	1.29	290.13	38.56	0.15
BA	72298.93	260	18.31	26.36	26.38	0.000157	0.75	488.68	60.81	0.08
BA	71563.65	260	18.28	26.32	26.34	0.00015	0.73	506.63	63.55	0.08
BA	71163.23	260	18.05	26.31	26.33	0.000134	0.67	526.44	71.68	0.08
BA	70592.32	260	16.76	26.24	26.29	0.000337	1.18	335.12	41.2	0.12
BA	70153.8	260	16.69	26.2	26.24	0.000323	1.16	338.47	40.1	0.12
BA	69869.56	260	16.6	26.19	26.22	0.000189	0.92	416.34	44.86	0.09
BA	68850.88	260	16.44	26.08	26.13	0.000358	1.18	331.24	36.98	0.13
BA	68097.29	260	16.99	26.02	26.06	0.000277	1.06	363.11	40.8	0.11
BA	67237.16	260	15.42	25.93	25.98	0.000278	1.16	330.76	33.17	0.12
BA	67071.61	260	16.76	25.87	25.96	0.00059	1.49	242.51	28.83	0.16
BA	66712.72	260	17.31	25.74	25.87	0.001172	2.05	204.28	25.6	0.23
BA	66502.64	260	15.73	25.77	25.81	0.000285	1.15	352.94	36.16	0.12
BA	66349.34	260	15.24	25.77	25.8	0.000182	0.93	424.87	45.17	0.09
BA	65891.16	260	16.18	25.75	25.77	0.000168	0.84	452.3	47.31	0.09
BA	65613.67	260	14.17	25.7	25.75	0.000297	1.25	331.53	31.88	0.12
BA	65454.99	260	17.1	25.67	25.73	0.000488	1.34	292.43	35.92	0.15
BA	65122.69	260	16.76	25.57	25.66	0.001018	1.23	225.96	27.77	0.13
BA	64758.91	260	16.76	25.4	25.53	0.001499	1.66	191.83	23.55	0.18
BA	63858	260	16.76	25.06	25.16	0.001254	1.47	213.77	28.65	0.17
BA	63293.18	260	15.24	25.06	25.08	0.000131	0.76	510.14	57.44	0.08
BA	62726.79	260	15.37	25.03	25.06	0.000192	0.92	419.62	46.74	0.1
BA	61091.66	260	15.24	24.94	24.97	0.000178	0.89	443.9	54.69	0.09
BA	60491.35	260	15.98	24.92	24.94	0.0001	0.62	582.59	67.8	0.07
BA	59729.88	260	15.24	24.86	24.9	0.000303	1.13	354.48	43.39	0.12
BA	59310.32	260	16.77	24.69	24.82	0.001075	1.81	198.5	29.02	0.21
BA	58816.8	260	16.33	24.69	24.72	0.000262	0.96	393.81	49.67	0.11
BA	58482.47	260	16.7	24.63	24.69	0.000428	1.2	313.83	40.09	0.14
BA	57985.86	260	16.62	24.6	24.63	0.00027	0.97	395.66	54.68	0.11
BA	56758.19	260	13.83	24.55	24.56	0.000108	0.74	526.18	55.47	0.07
BA	54951.46	260	15.66	24.45	24.48	0.000232	0.95	411.2	50.75	0.1
BA	53987.75	260	14.32	24.31	24.38	0.000531	1.57	272.34	28.48	0.16
BA	53252.57	260	16.76	23.97	24.15	0.002547	1.88	162.58	26.55	0.22
BA	52870.24	260	14.73	24.02	24.04	0.0002	0.85	495.19	91.57	0.09
BA	52275.16	260	15.65	23.95	23.99	0.000324	1.07	362.35	47.18	0.12
BA	51019	260	17.24	23.77	23.82	0.000701	1.18	303.07	58.4	0.16

BA	50191.88	260	15.24	23.48	23.58	0.001202	1.37	212.11	25.75	0.15
BA	49724.13	260	13.92	23.51	23.53	0.000078	0.58	654.62	75.32	0.06
BA	47526.98	260	14.73	23.43	23.45	0.000166	0.79	474.9	56.61	0.09
BA	46916.46	260	18.08	23.33	23.39	0.000742	1.21	282.01	51.04	0.17
BA	46287.4	260	14.44	23.33	23.34	0.000091	0.59	616.11	71.49	0.06
BA	44984.53	260	14.07	23.28	23.29	0.000138	0.74	512.07	61.01	0.08
BA	43722.66	260	14.18	23.21	23.23	0.000194	0.88	436.34	49.79	0.09
BA	43180.79	260	14.81	23.16	23.19	0.00028	1.01	376.68	45.51	0.11
BA	42561.23	260	13.72	22.95	23.08	0.001475	1.7	190.98	24.07	0.18
BA	41790.33	260	13.54	22.78	22.84	0.000584	1.29	283.81	34.12	0.15
BA	41011	260	13.83	22.76	22.77	0.000131	0.73	528.93	63.65	0.08
BA	39834.5	260	14	22.69	22.72	0.000209	0.9	430.27	51.52	0.1
BA	38438.09	260	13.72	22.53	22.58	0.000511	0.95	296.84	33.67	0.1
BA	37849.79	260	13.85	22.29	22.45	0.001129	2.02	194.74	24.41	0.23
BA	37376.59	260	16.11	22.3	22.33	0.000305	0.84	397.85	56.93	0.11
BA	36473.56	260	13.18	22.13	22.2	0.000732	1.12	255.35	28.62	0.12
BA	35723.82	260	13.72	22.07	22.1	0.000257	0.91	421.44	59.76	0.1
BA	35299.99	260	14.27	21.99	22.05	0.000572	1.33	293.84	39.43	0.16
BA	33843.18	260	13.72	21.61	21.69	0.001002	1.28	229.63	29.1	0.15
BA	33084.58	260	12.19	21.55	21.58	0.000216	0.96	404.22	45.97	0.1
BA	32027	260	13.69	21.47	21.5	0.000254	0.57	394.84	46.03	0.07
BA	30751.24	260	13.72	21.36	21.39	0.000321	0.71	374.9	48.88	0.08
BA	29591.09	260	14.92	21.24	21.28	0.000341	0.92	375.36	54.93	0.12
BA	28928.12	260	12.23	21.2	21.23	0.000167	0.83	456.13	51.04	0.09
BA	27743.95	260	14.83	21.06	21.12	0.000619	1.19	289.07	43.61	0.16
BA	27053.15	260	13.34	20.99	21.03	0.000293	0.94	381.19	52.95	0.11
BA	25510.07	260	14.21	20.63	20.75	0.001468	1.81	205.01	34.19	0.24
BA	24987.13	260	11.93	20.48	20.55	0.000826	1.2	245.3	29.22	0.13
BA	24381.15	260	11.84	20.36	20.42	0.000636	1.09	271.09	32.03	0.12
BA	22038.99	260	11.1	20.11	20.14	0.000238	0.92	398.91	49.01	0.1
BA	18648.79	260	12.72	19.85	19.88	0.000251	0.82	433.65	64.48	0.1
BA	17757.68	260	10.29	19.75	19.8	0.000366	1.25	318.07	34.53	0.13
BA	16496.04	260	12.19	19.27	19.48	0.002927	1.8	151.59	21.41	0.22
BA	14247.63	260	13.18	17.91	18.01	0.001549	1.61	223.08	48.33	0.24
BA	12789.76	260	11.95	17.59	17.63	0.000518	0.96	332.05	58.34	0.14
BA	10641.71	260	10.7	17.23	17.28	0.000539	1.2	310.93	50.64	0.15
BA	8383.456	260	7.99	17.01	17.04	0.00022	0.91	410.83	50.38	0.1
BA	6609.535	260	10.81	16.58	16.72	0.001872	1.94	183.53	32.82	0.27
BA	5180.641	260	7.18	16.47	16.49	0.000214	0.92	419.06	48.91	0.1
BA	3616.872	260	9.17	16.16	16.28	0.001273	1.83	205.77	30.53	0.23
BA	2623.655	260	6.62	16.09	16.12	0.000226	0.93	411.24	51.37	0.1
BA	700.579	260	9.25	15.85	15.91	0.000618	1.28	288.93	46.52	0.16
BA	190.2835	260	12.41	15.09	15.59	0.017026	3.09	101.04	45.55	0.69

Appendix 8.2. Parameters calculated in HEC-RAS during the simulation for a flow rate of 650 m /s.^{3}

Reach	River Sta	Q Total	Min Ch El	W.S. Elev	E.G. Elev	E.G. Slope	Vel Chnl	Flow Area	Top Width	Froude # Chl
		(m3/s)	(m)	(m)	(m)	(m/m)	(m/s)	(m2)	(m)	
BA	83848.21	650	19.81	32.15	32.23	0.000337	1.41	549.64	50.27	0.13
BA	82974.85	650	21.01	32.07	32.13	0.000362	1.34	571.44	67.45	0.13
BA	82439.9	650	19.86	32.01	32.07	0.000322	1.4	572.67	47.88	0.13
BA	81234.16	650	19.86	31.84	31.92	0.000429	1.6	496.73	42.55	0.15
BA	80343.55	650	17.57	31.76	31.82	0.000266	1.37	600.37	47.51	0.12
BA	80182.37	650	18.42	31.76	31.8	0.00022	1.19	666.58	55.25	0.11
BA	78705.05	650	19.71	31.61	31.68	0.000372	1.47	546.42	47.47	0.14
BA	78479.98	650	19.36	31.55	31.65	0.000432	1.62	491.38	41.51	0.15

BA	77914.52	650	19.43	31.52	31.57	0.000299	1.3	597.14	50.82	0.12
BA	77400.86	650	16.76	31.36	31.5	0.000616	2.04	414.24	32.23	0.18
BA	77180.63	650	12.73	31.4	31.45	0.000183	1.28	642.67	38.45	0.1
BA	76655.31	650	18.31	31.34	31.41	0.000354	1.54	546.4	43.72	0.14
BA	75371.32	650	17.02	31.25	31.3	0.00021	1.24	674.48	53.53	0.11
BA	75059.63	650	19.81	31.18	31.26	0.000601	1.27	465.6	40.63	0.12
BA	74031.25	650	16.76	31.17	31.19	0.000088	0.81	959.88	74.71	0.07
BA	72786.73	650	18.57	31.01	31.11	0.000513	1.73	469.17	38.56	0.16
BA	72298.93	650	18.31	31.02	31.05	0.000155	1	772.29	60.81	0.09
BA	71563.65	650	18.28	30.99	31.02	0.000148	0.98	803.11	63.55	0.09
BA	71163.23	650	18.05	30.98	31	0.000121	0.87	861.11	71.68	0.08
BA	70592.32	650	16.76	30.88	30.96	0.000352	1.58	526.38	41.2	0.14
BA	70153.8	650	16.69	30.83	30.92	0.000349	1.58	524.42	40.1	0.14
BA	69869.56	650	16.6	30.83	30.88	0.000223	1.3	624.51	44.86	0.11
BA	68850.88	650	16.44	30.7	30.78	0.00042	1.7	501.75	36.98	0.15
BA	68097.29	650	16.99	30.62	30.7	0.000316	1.49	551.01	40.8	0.13
BA	67237.16	650	15.42	30.5	30.61	0.000352	1.67	482.3	33.17	0.14
BA	67071.61	650	16.76	30.42	30.58	0.000615	2.03	373.6	28.83	0.18
BA	66712.72	650	17.31	30.24	30.48	0.001245	2.83	319.51	25.6	0.25
BA	66502.64	650	15.73	30.31	30.39	0.000369	1.68	517.1	36.16	0.14
BA	66349.34	650	15.24	30.32	30.37	0.000226	1.33	630.35	45.17	0.11
BA	65891.16	650	16.18	30.29	30.34	0.000212	1.23	667.37	47.31	0.11
BA	65613.67	650	14.17	30.2	30.31	0.000418	1.87	474.99	31.88	0.15
BA	65454.99	650	17.1	30.18	30.29	0.000528	1.86	454.29	35.92	0.17
BA	65122.69	650	16.76	30.06	30.21	0.001189	1.49	350.48	27.77	0.13
BA	64758.91	650	16.76	29.84	30.05	0.001851	2.13	296.26	23.55	0.19
BA	63858	650	16.76	29.46	29.62	0.001361	1.82	339.79	28.65	0.17
BA	63293.18	650	15.24	29.48	29.52	0.000155	1.07	763.96	57.44	0.09
BA	62726.79	650	15.37	29.43	29.49	0.000232	1.3	625.42	46.74	0.11
BA	61091.66	650	15.24	29.33	29.38	0.000194	1.19	684.26	54.69	0.1
BA	60491.35	650	15.98	29.32	29.35	0.000113	0.87	880.73	67.8	0.08
BA	59729.88	650	15.24	29.23	29.3	0.000341	1.56	544.05	43.39	0.13
BA	59310.32	650	16.77	29.01	29.22	0.000938	2.31	323.85	29.02	0.22
BA	58816.8	650	16.33	29.05	29.1	0.000284	1.33	610.2	49.67	0.12
BA	58482.47	650	16.7	28.97	29.06	0.00045	1.66	487.53	40.09	0.15
BA	57985.86	650	16.62	28.95	29	0.000261	1.27	633.46	54.68	0.12
BA	56758.19	650	13.83	28.89	28.92	0.000139	1.05	766.98	55.47	0.09
BA	54951.46	650	15.66	28.77	28.82	0.000257	1.31	630.38	50.75	0.12
BA	53987.75	650	14.32	28.54	28.69	0.000735	2.35	393.03	28.48	0.2
BA	53252.57	650	16.76	28.17	28.42	0.002316	2.08	274.05	26.55	0.2
BA	52870.24	650	14.73	28.28	28.31	0.000144	0.94	885.7	91.57	0.08
BA	52275.16	650	15.65	28.2	28.26	0.000354	1.49	562.8	47.18	0.13
BA	51019	650	17.24	28.06	28.12	0.000446	1.38	553.46	58.4	0.14
BA	50191.88	650	15.24	27.73	27.91	0.001524	1.76	321.6	25.75	0.16
BA	49724.13	650	13.92	27.81	27.83	0.000092	0.81	978.03	75.32	0.07
BA	47526.98	650	14.73	27.7	27.74	0.000192	1.12	716.78	56.61	0.1
BA	46916.46	650	18.08	27.61	27.68	0.000515	1.51	500.09	51.04	0.16
BA	46287.4	650	14.44	27.6	27.63	0.000107	0.83	921.97	71.49	0.07
BA	44984.53	650	14.07	27.54	27.58	0.000159	1.04	772.37	61.01	0.09
BA	43722.66	650	14.18	27.45	27.5	0.000239	1.27	647.72	49.79	0.11
BA	43180.79	650	14.81	27.39	27.45	0.000325	1.44	569.17	45.51	0.13
BA	42561.23	650	13.72	27.09	27.32	0.001884	2.23	290.64	24.07	0.2
BA	41790.33	650	13.54	26.9	27.01	0.000725	1.92	424.19	34.12	0.18
BA	41011	650	13.83	26.89	26.92	0.000154	1.01	791.84	63.65	0.09
BA	39834.5	650	14	26.8	26.85	0.000251	1.28	642.04	51.52	0.11
BA	38438.09	650	13.72	26.58	26.68	0.000698	1.26	433.22	33.67	0.11
BA	37849.79	650	13.85	26.2	26.5	0.001337	2.85	289.94	24.41	0.26
BA	37376.59	650	16.11	26.29	26.34	0.000306	1.19	624.84	56.93	0.12

BA	36473.56	650	13.18	26.05	26.18	0.001066	1.53	367.44	28.62	0.14
BA	35723.82	650	13.72	26.01	26.05	0.000268	1.23	656.56	59.76	0.11
BA	35299.99	650	14.27	25.89	25.99	0.000651	1.89	447.63	39.43	0.18
BA	33843.18	650	13.72	25.4	25.56	0.001327	1.69	340.12	29.1	0.16
BA	33084.58	650	12.19	25.34	25.4	0.000298	1.41	578.39	45.97	0.13
BA	32027	650	13.69	25.24	25.3	0.000344	0.75	568.13	46.03	0.07
BA	30751.24	650	13.72	25.09	25.15	0.00039	0.91	557.44	48.88	0.09
BA	29591.09	650	14.92	24.96	25.02	0.00036	1.3	579.79	54.93	0.13
BA	28928.12	650	12.23	24.91	24.96	0.000234	1.23	645.47	51.04	0.11
BA	27743.95	650	14.83	24.73	24.83	0.00066	1.7	448.92	43.61	0.18
BA	27053.15	650	13.34	24.66	24.72	0.000339	1.33	575.25	52.95	0.13
BA	25510.07	650	14.21	24.22	24.42	0.001381	2.43	327.51	34.19	0.25
BA	24987.13	650	11.93	24.03	24.18	0.001231	1.65	348.99	29.22	0.15
BA	24381.15	650	11.84	23.85	23.97	0.00096	1.53	382.9	32.03	0.14
BA	22038.99	650	11.1	23.5	23.57	0.000338	1.39	565.15	49.01	0.13
BA	18648.79	650	12.72	23.18	23.23	0.000291	1.16	648.18	64.48	0.12
BA	17757.68	650	10.29	22.99	23.12	0.00059	1.95	430.15	34.53	0.18
BA	16496.04	650	12.19	22.19	22.59	0.004494	2.46	213.92	21.41	0.25
BA	14247.63	650	13.18	20.8	20.94	0.001366	2.09	362.52	48.33	0.24
BA	12789.76	650	11.95	20.47	20.55	0.000597	1.4	500.28	58.34	0.16
BA	10641.71	650	10.7	20.02	20.12	0.000686	1.71	452.34	50.64	0.18
BA	8383.456	650	7.99	19.69	19.76	0.000375	1.42	545.74	50.38	0.14
BA	6609.535	650	10.81	18.95	19.25	0.002519	2.88	261.42	32.82	0.33
BA	5180.641	650	7.18	18.77	18.83	0.000423	1.51	531.5	48.91	0.14
BA	3616.872	650	9.17	18.12	18.41	0.002361	2.97	265.68	30.53	0.33
BA	2623.655	650	6.62	18	18.08	0.000474	1.53	509.47	51.37	0.15
BA	700.579	650	9.25	17.51	17.65	0.001202	2.08	365.92	46.52	0.23
BA	190.2835	650	12.41	16.37	17.13	0.017002	4.27	159.38	45.55	0.75

Appendix 8.3. Parameters calculated in HEC-RAS during the simulation for a flow rate of 780 m /s.^{3}

Reach	River Sta	Q Total	Min Ch El	W.S. Elev	E.G. Elev	E.G. Slope	Vel Chnl	Flow Area	Top Width	Froude # Chl
		(m3/s)	(m)	(m)	(m)	(m/m)	(m/s)	(m2)	(m)	
BA	83848.21	780	19.81	34.15	34.23	0.000319	1.52	650.12	50.27	0.13
BA	82974.85	780	21.01	34.08	34.15	0.000301	1.37	707.41	67.45	0.12
BA	82439.9	780	19.86	34.02	34.1	0.000318	1.54	669	47.88	0.13
BA	81234.16	780	19.86	33.85	33.95	0.000417	1.75	582.25	42.55	0.15
BA	80343.55	780	17.57	33.77	33.85	0.00027	1.51	696.01	47.51	0.12
BA	80182.37	780	18.42	33.77	33.83	0.000216	1.3	777.93	55.25	0.11
BA	78705.05	780	19.71	33.62	33.7	0.00036	1.61	642.08	47.47	0.14
BA	78479.98	780	19.36	33.56	33.67	0.000422	1.78	574.79	41.51	0.15
BA	77914.52	780	19.43	33.53	33.6	0.000293	1.43	699.61	50.82	0.12
BA	77400.86	780	16.76	33.35	33.52	0.00063	2.27	478.46	32.23	0.18
BA	77180.63	780	12.73	33.4	33.47	0.000207	1.47	719.61	38.45	0.11
BA	76655.31	780	18.31	33.34	33.43	0.00036	1.71	633.88	43.72	0.14
BA	75371.32	780	17.02	33.25	33.31	0.000212	1.36	781.67	53.53	0.11
BA	75059.63	780	19.81	33.18	33.28	0.000595	1.34	546.84	40.63	0.12
BA	74031.25	780	16.76	33.17	33.2	0.000088	0.89	1109.61	74.71	0.07
BA	72786.73	780	18.57	33	33.12	0.000509	1.92	545.76	38.56	0.16
BA	72298.93	780	18.31	33.02	33.06	0.000156	1.11	893.67	60.81	0.09
BA	71563.65	780	18.28	32.99	33.02	0.000149	1.08	929.97	63.55	0.09
BA	71163.23	780	18.05	32.97	33	0.00012	0.96	1004.29	71.68	0.08
BA	70592.32	780	16.76	32.86	32.96	0.00036	1.75	608.1	41.2	0.14
BA	70153.8	780	16.69	32.81	32.92	0.000359	1.75	603.84	40.1	0.14
BA	69869.56	780	16.6	32.81	32.88	0.000234	1.45	713.46	44.86	0.11

BA	68850.88	780	16.44	32.67	32.78	0.000442	1.91	574.6	36.98	0.16
BA	68097.29	780	16.99	32.59	32.69	0.000329	1.67	631.35	40.8	0.14
BA	67237.16	780	15.42	32.45	32.59	0.000374	1.87	547.08	33.17	0.15
BA	67071.61	780	16.76	32.36	32.56	0.000624	2.24	429.65	28.83	0.18
BA	66712.72	780	17.31	32.16	32.46	0.001281	3.15	368.66	25.6	0.26
BA	66502.64	780	15.73	32.25	32.36	0.000398	1.9	587.26	36.16	0.15
BA	66349.34	780	15.24	32.26	32.33	0.000241	1.49	718.24	45.17	0.12
BA	65891.16	780	16.18	32.24	32.29	0.000227	1.39	759.41	47.31	0.11
BA	65613.67	780	14.17	32.12	32.26	0.00046	2.12	536.25	31.88	0.16
BA	65454.99	780	17.1	32.1	32.24	0.000546	2.08	523.42	35.92	0.17
BA	65122.69	780	16.76	31.97	32.15	0.001265	1.59	403.71	27.77	0.13
BA	64758.91	780	16.76	31.73	31.98	0.002004	2.3	340.9	23.55	0.19
BA	63858	780	16.76	31.34	31.52	0.001426	1.96	393.58	28.65	0.17
BA	63293.18	780	15.24	31.37	31.41	0.000163	1.19	872.33	57.44	0.1
BA	62726.79	780	15.37	31.31	31.38	0.000245	1.46	713.27	46.74	0.12
BA	61091.66	780	15.24	31.21	31.27	0.0002	1.31	786.95	54.69	0.11
BA	60491.35	780	15.98	31.2	31.23	0.000118	0.97	1008.12	67.8	0.08
BA	59729.88	780	15.24	31.09	31.18	0.000356	1.73	625	43.39	0.14
BA	59310.32	780	16.77	30.85	31.1	0.000916	2.52	377.27	29.02	0.22
BA	58816.8	780	16.33	30.91	30.97	0.000294	1.49	702.52	49.67	0.13
BA	58482.47	780	16.7	30.81	30.93	0.000461	1.84	561.57	40.09	0.16
BA	57985.86	780	16.62	30.8	30.86	0.000263	1.4	734.85	54.68	0.12
BA	56758.19	780	13.83	30.74	30.78	0.000149	1.18	869.67	55.47	0.09
BA	54951.46	780	15.66	30.61	30.67	0.000267	1.46	723.91	50.75	0.12
BA	53987.75	780	14.32	30.35	30.54	0.00081	2.67	444.4	28.48	0.21
BA	53252.57	780	16.76	29.96	30.24	0.002334	2.18	321.63	26.55	0.19
BA	52870.24	780	14.73	30.1	30.13	0.000134	1	1052.12	91.57	0.08
BA	52275.16	780	15.65	30.01	30.08	0.000368	1.66	648.12	47.18	0.14
BA	51019	780	17.24	29.87	29.94	0.000413	1.49	659.44	58.4	0.14
BA	50191.88	780	15.24	29.52	29.73	0.001666	1.91	367.6	25.75	0.16
BA	49724.13	780	13.92	29.61	29.64	0.000097	0.91	1114.08	75.32	0.07
BA	47526.98	780	14.73	29.5	29.55	0.000203	1.25	818.52	56.61	0.11
BA	46916.46	780	18.08	29.4	29.49	0.000487	1.64	591.65	51.04	0.16
BA	46287.4	780	14.44	29.4	29.43	0.000113	0.94	1050.46	71.49	0.08
BA	44984.53	780	14.07	29.33	29.37	0.000168	1.16	881.7	61.01	0.1
BA	43722.66	780	14.18	29.23	29.29	0.000256	1.43	736.48	49.79	0.12
BA	43180.79	780	14.81	29.16	29.24	0.000343	1.61	649.97	45.51	0.14
BA	42561.23	780	13.72	28.83	29.09	0.002067	2.44	332.42	24.07	0.2
BA	41790.33	780	13.54	28.62	28.76	0.00078	2.18	483.05	34.12	0.19
BA	41011	780	13.83	28.62	28.66	0.000162	1.13	902.27	63.65	0.09
BA	39834.5	780	14	28.53	28.59	0.000267	1.44	730.96	51.52	0.12
BA	38438.09	780	13.72	28.28	28.41	0.000774	1.37	490.49	33.67	0.11
BA	37849.79	780	13.85	27.82	28.2	0.001414	3.19	329.66	24.41	0.27
BA	37376.59	780	16.11	27.96	28.02	0.000312	1.33	719.99	56.93	0.13
BA	36473.56	780	13.18	27.68	27.85	0.001204	1.68	414.35	28.62	0.14
BA	35723.82	780	13.72	27.65	27.71	0.000275	1.36	755.06	59.76	0.12
BA	35299.99	780	14.27	27.52	27.65	0.000686	2.12	511.94	39.43	0.19
BA	33843.18	780	13.72	26.99	27.18	0.001464	1.85	386.33	29.1	0.16
BA	33084.58	780	12.19	26.92	27	0.000329	1.6	651.3	45.97	0.13
BA	32027	780	13.69	26.82	26.89	0.000378	0.81	640.75	46.03	0.07
BA	30751.24	780	13.72	26.66	26.73	0.000417	0.97	634.01	48.88	0.09
BA	29591.09	780	14.92	26.53	26.6	0.000371	1.45	665.52	54.93	0.14
BA	28928.12	780	12.23	26.47	26.53	0.000259	1.4	724.89	51.04	0.12
BA	27743.95	780	14.83	26.27	26.38	0.000683	1.92	515.99	43.61	0.18
BA	27053.15	780	13.34	26.2	26.27	0.000357	1.49	656.7	52.95	0.14
BA	25510.07	780	14.21	25.71	25.96	0.001386	2.69	378.71	34.19	0.26
BA	24987.13	780	11.93	25.51	25.7	0.001397	1.82	392.39	29.22	0.16
BA	24381.15	780	11.84	25.31	25.47	0.001092	1.69	429.73	32.03	0.15

BA	22038.99	780	11.1	24.93	25.01	0.000375	1.58	635.04	49.01	0.14
BA	18648.79	780	12.72	24.58	24.64	0.000306	1.3	738.71	64.48	0.12
BA	17757.68	780	10.29	24.36	24.52	0.000677	2.23	477.26	34.53	0.19
BA	16496.04	780	12.19	23.41	23.92	0.005135	2.71	240.12	21.41	0.26
BA	14247.63	780	13.18	22.01	22.18	0.001348	2.29	421.27	48.33	0.25
BA	12789.76	780	11.95	21.68	21.78	0.000625	1.58	571.06	58.34	0.17
BA	10641.71	780	10.7	21.2	21.32	0.000738	1.93	512.03	50.64	0.19
BA	8383.456	780	7.99	20.83	20.91	0.000435	1.64	603.01	50.38	0.15
BA	6609.535	780	10.81	19.96	20.33	0.002735	3.26	294.58	32.82	0.35
BA	5180.641	780	7.18	19.75	19.84	0.000513	1.76	579.78	48.91	0.16
BA	3616.872	780	9.17	18.95	19.33	0.002814	3.45	290.98	30.53	0.36
BA	2623.655	780	6.62	18.82	18.92	0.000585	1.79	551.3	51.37	0.17
BA	700.579	780	9.25	18.21	18.4	0.001453	2.42	398.37	46.52	0.26
BA	190.2835	780	12.41	16.92	17.8	0.017015	4.72	184.09	45.55	0.77

Printed by Books on Demand GmbH, Norderstedt / Germany